THE BOTANISTS

H. C. Watson, Vice-President of the Botanical Society of London and in whose memory the present Society has named its journal *Watsonia*. Oil painting by Margaret Sarah Carpenter, subscribed for by the members in 1846.

THE
BOTANISTS

A history of the Botanical Society of the British Isles through a hundred and fifty years

BY

DAVID ELLISTON ALLEN

ST PAUL'S BIBLIOGRAPHIES

1986

First published by St Paul's Bibliographies,
West End House, 1 Step Terrace, Winchester, U.K.,
in 1986.

© David Elliston Allen 1986

British Library Cataloguing in Publication Data
Allen, David Elliston
　　The botanists: the Botanical Society of the
　British Isles through a hundred and fifty years.
　　1. Botanical Society of the British Isles—
　History
I. Title
581'.06041　　　　QK1

ISBN 0-906795-36-2

Printed and bound in Great Britain by
Redwood Burn Limited
Trowbridge

As Patron of the Botanical Society of the British Isles, I join with all the Members in celebrating the sesquicentenary. It is a remarkable milestone, and to have reached it reflects not only the scholarship and enthusiasm of so many people over so many years, but also the value placed on the work of the Society. The fact that the membership is larger than it ever has been and that the scientific publications enjoy an international reputation are sources of much satisfaction.

We must now look to the future and how best to continue to serve a branch of learning which has brought together amateurs and professionals and taught them to enjoy the flora of the countryside.

Clarence House   ELIZABETH R
S.W.1.   Queen Mother
20th September 1985

CONTENTS

Foreword	*Page*	xi
Preface		xiii

PART ONE
The Botanical Society of London

1. The origins		3
2. A society is born		11
3. Watson takes over		25
4. The other side		39
5. Collapse		52

PART TWO
The Botanical Exchange Club

6. Interlude at Thirsk		69
7. The years of obscurity		77
8. Civil war		91
9. King Druce		105
10. Return to democracy		116

PART THREE
The Botanical Society of the British Isles

11. The great efflorescence		133
12. Full steam ahead		148
13. Into the future		159
Notes		175

APPENDICES

1. Membership figures of the Botanical Society of London		201
2. The Members of the Botanical Society of London		202
3. Officers 1836–1986		222
4. Principal BSBI conferences		225
Index		227

LIST OF ILLUSTRATIONS

1. H. C. Watson: portrait by Margaret Sarah Carpenter   *Frontispiece*
   Courtesy of British Museum (Natural History)
2. J. E. Gray: portrait by Margaret Sarah Carpenter   *Facing page*   22
   Courtesy of British Museum (Natural History)
3. 20 Bedford Street   *Between pages*   58/60
   Photograph by V. Fleming
4. Title page of annual report 1838   ,,
   Courtesy of Merseyside County Museums
5. Page from the auction catalogue of 1857   ,,
   Courtesy of British Museum (Natural History)
6. J. G. Baker   *Facing page*   71
   Courtesy of British Museum (Natural History)
7. E. S. Marshall   87
   Courtesy of British Museum (Natural History)
8. G. Claridge Druce: portrait by P. A. de Laszlo   105
   Courtesy of Bodleian Library
9. James Britten   *Between pages*   121/3
   Courtesy of British Museum (Natural History)
10. C. E. Moss   ,,
    Courtesy of Botany School, Cambridge
11. W. H. Pearsall   ,,
    Courtesy of A. W. H. Pearsall and Cumbria Record Office
12. G. C. Druce examining Loddon Pondweed   *Facing page*   134
13. Loddon Pondweed   137
    Courtesy of Stella Ross-Craig and G. Bell & Sons
14. Miss M. S. Campbell   142
15. A. J. Wilmott and Francis Druce   142
    Courtesy of British Museum (Natural History)
16. F. H. Perring, with map printer   *Between pages*   154/7
17. Distribution Map: National Grid   ,,

18. Field card *Between pages* 154/7
19. Primrose: water-colour by William Kilburn "
    Courtesy of British Museum (Natural History)
20. The Queen Mother leading a party from seeing the Oyster
    Plant *Facing page* 160
    Courtesy of M. Luciani
21. Oyster Plant: engraving by Dillenius 163
    Courtesy of British Museum (Natural History)
22. Design for the cover of the abortive
    'The Botanist' *Between pages* 164/6
    Photograph by V. Fleming
23. The post-War rise in membership "
    Courtesy of J. G. Dony
24. J. E. Lousley, the architect of the membership rise "
    Courtesy of Mrs C. M. Dony
25. Military Orchid: water-colour by E. J. Bedford *Facing page* 172
    Courtesy of British Museum (Natural History)

The seal of the Botanical Society of London, illustrated on the Part One half title, is taken from a manuscript in the property of the Royal Botanic Gardens, Kew, and is reproduced by their kind permission.

FOREWORD

Anniversaries, both public and private, are often seen as opportunities for reflection on the years that have passed, modest gratification at the progress that has been made and contemplation of the lessons that should be learnt for the future. The Botanical Society of the British Isles is no exception, and one of the first decisions made by Council in connection with the Society's sesquicentenary in 1986, was to commission a History of the Society, so that all members and the world at large could learn more about its origins and give informed thought to its future. The BSBI is unusually fortunate in that its official historian, Mr David Allen, a long-standing member with a proven record of involvement in many aspects of the Society's activities, is also a distinguished authority on the development of natural history in Britain, with special reference to its relationship with social change over the years. It is highly appropriate that his many and varied contributions have been recognised by his election for the sesquicentennial period as the Society's President, and it is a great pleasure for me, as immediate Past President, to have the opportunity of introducing this account of the first 150 years of the BSBI. While the text is scholarly in the best traditions of social and natural history, nobody should be deterred from reading it by the fear that it might be dull or boring for the layman. On the contrary, its pages are filled with incident and anecdotes of some extraordinary personalities, under whom the Society has prospered or, in some cases, survived. One can only hope that the present membership will provide as rich a tapestry, albeit with less drama, for the contemplation of our successors.

We believe that it is right, at this time, to take pride in the scientific achievements of the Society, most especially in its publications, which are assured of their rightful place in the international literature of science, and which contribute significantly to our understanding of plant diversity and distribution. Natural history in the British Isles has always been characterised by a remarkably close association between amateurs and professionals, and this has long been admired by naturalists from other countries. The BSBI's strength over the years has often been attributed to this phenomenon. The apogee to date of the Society's scientific achievement was undoubtedly the Distribution Maps Scheme which culminated in the publication of the *Atlas of the British Flora*. This was based on mechanised mapping techniques which, in the computer age, now have a dinosaur-like air but which, at that time, led the world in biological mapping. The British genius for amateur–professional cooperation lay at the very heart of that project and, through it, hundreds of members were profitably engaged in a project which achieved a significance greater than the sum of the individual efforts involved. This, coupled with another British talent for making do with modest resources, truly amazed a very distinguished American academic, whom I took on a visit to the Maps Scheme office at Cambridge at the height of its operation. He was astonished to see the small attic office and the cramped cellar in which the card sorting and mapping machines were housed. His comment that "in our country the first move would have been to erect a multi-million dollar building, followed by the appointment of a substantial professional staff", spoke volumes for his admiration for a major scientific and organisational achievement, attained in a characteristically modest manner, by a Society that is also remarkable for the way in which Britain and the Republic of Ireland have pooled their botanical talent in productive harmony.

Although the Society is celebrating its first 150 years, Mr Allen has thought it appropriate, with a historian's discretion, to conclude his story at about 1965. The BSBI now has 2,700 members with a wide range of activities and it will be for the next historian of the Society to comment on our own times, with the advantage of hindsight and a broader perspective than we can currently envisage.

Finally, it is a special privilege to acknowledge the message to the Society from our Patron, Her Majesty Queen Elizabeth, the Queen Mother, on the occasion of our Sesquicentenary, which is reproduced in these introductory pages. Her very real love of plants both in gardens and in the countryside is common knowledge, but her hospitality both in the field and at home, on the occasion of one of the Society's field meetings in Scotland, is known in detail only to those few members who so greatly enjoyed the occasion.

John F. M. Cannon
President 1983–85

PREFACE

It is more than seventy years now since the publishing of a full-scale history of the Society was first suggested. This was in the spring of 1914, when, with the eightieth anniversary in prospect, the then Honorary Secretary, G. C. Druce, took the opportunity of one of his circulars to members to invite donations towards the cost. The response, however, proved so disappointing that the idea had to be dropped. In the early 1930s, with the centenary looming, it must have been raised again; but in the event the most that appeared on that occasion were two rather slight accounts of the earlier periods, contributed to one of the Reports.

The late J. E. Lousley had long had in mind making good this rather shaming deficiency and, by way of a preliminary, published briefly but valuably on the subject in his later years. When, after his sudden death in 1976, I agreed to take over the task, I did so in the expectation of inheriting a substantial accumulation of material from him. To my surprise, however, this did not prove to be the case: sadly, almost all of his undoubtedly extensive knowledge of the Society's past had failed to be committed to paper. In effect, therefore, I had to begin afresh.

The task, I was all too aware, was not going to be easy. The modern Society's records were believed to date only from 1941, everything that existed before then supposedly having been destroyed in the bombing of the then Honorary Treasurer's London flat (though much of the official correspondence of the 1930s, it later emerged, had survived among the Druce papers at Oxford). Of the ancestral body, the Botanical Society of London, similarly, no records survived apart from one solitary minute-book, which, as I already knew, was disappointingly uninformative – for it did not extend to the all-important Council meetings. In so far as it was going to be possible to reconstruct events, I would therefore have to depend on what had been reported in print and on such letters and other unpublished documents as I could manage to locate.

The paucity of information about the Botanical Society of London in particular, which flourished at a period of especial interest to historians of science, seemed to make that body an obvious candidate for the comparatively novel technique of 'prosopography' (or social analysis derived from the building-up of numerous individual biographies); and it was most opportune, therefore, that at that point I had the good fortune to be invited by Dr (now Professor) Roy MacLeod to spend a term at the University of Sussex as a Visiting Fellow in the History and Social Studies of Science Area. With the aid of a research grant awarded by the Leverhulme Trust, for which I would here like to express my gratitude, this enabled me to undertake the laborious groundwork which a prosopographical study necessarily entails and which otherwise I might not have been emboldened to embark upon. It further gave me the chance to go through the fifty or more boxes of unlisted papers which constitute the Druce Archive, the key source for the Society's history between 1900 and 1935. I would specially like to thank Mrs H. J. McArdle, then Librarian of the Oxford Botany School, and her colleagues for all the assistance they gave me during those memorably arduous two weeks.

To the many other librarians, archivists and curators of museums who have helped me in the course of the research for this book, more particularly in my attempts to identify the more elusive members of the Botanical Society of London, I should also like to take this opportunity of acknowledging my debt. They are far too numerous, alas, for me to name them all individually; I can only thank them collectively for their kindness in responding to my queries or granting me access to papers. Among them, even so, are several who can hardly not be singled out for mention, in view of the particular degree of trouble they went to in searching the collections in their care for BSL specimens and providing me with lengthy lists or print-outs. They are: Mrs P. J. Copson, of Warwick Museum (Herb. Perry), P. S. Davis, D. Kirkup and Mrs J. Hebron, of the Hancock Museum, Newcastle upon Tyne (Herb. Embleton, Herb. Bigge), E. G. Hancock and Miss P. Cottom, of Bolton Museum (Herb. Mason), Miss E. M. C. Isherwood and L. Brown, of Holmesdale Natural History Club Museum, Reigate (Herb. Brewer) and Mrs S. J. Patrick, of Derby City Museum (Herb. Whittaker). In this same connection too I need to thank more especially those institutions which permitted me to make comprehensive manual searches of my own of herbaria known to be rich in BSL material: the British Museum (Natural History) (Herb. Syme), the Castle Museum, Norwich (Herb. Woodward), the Horniman Museum (Herb. Brocas), the Royal Botanic Gardens, Kew (Herb. Watson) and the Yorkshire Museum (Herb. Hailstone). To Mrs B. D. Greenwood I also owe a deep debt for bringing to my attention the supremely valuable T. B. Hall papers in Merseyside County Museums and for supplying copies of these.

A number of members prominent in the Society's affairs in the past were good enough to grant me interviews and to submit to my questioning with helpfulness and patience. Foremost among these were the late Miss Campbell, John Chapple, John Dony, Douglas Kent and Sir George Taylor. John Dony and Douglas Kent placed me further in their debt, along with John Akeroyd, Mary Briggs, John Cannon, Arthur Chater, Chris Dony, Richard Gornall, Jack Morrell, Mike Walpole and David Webb, in agreeing to the onerous task of reading the book in draft. It goes without saying that much useful comment resulted but that they bear no responsibility for what has finally emerged. As Honorary Secretary of the Society's Publications Committee Arthur Chater has played a particularly central role: I am grateful to him for continuing help and advice. I would also specially like to thank Vaughan Fleming for assisting with the illustrations and Audrey Tester for her, as always, impeccable typescript. Finally, I must pay tribute to the endurance of my wife: it must have seemed at times as if Watson and Druce had taken up residence in our house. She has borne my obsession with fortitude and has been, unfailingly, my most unsparing critic.

D.E.A.

Winchester
October, 1985

PART ONE

The Botanical Society of London

CHAPTER 1

The Origins

As long as botanists have been numerous enough in one place to meet for an occasional gathering there have been botanical societies. Some believe they discern one even as early as the 1620s, in the shape of the *socii itinerantes* of Thomas Johnson, the group of young apothecaries which he regularly led out on pioneer 'simpling voyages' into the countryside of southern England. The earliest that we know of for certain, though – and probably the earliest in the world – flourished in Restoration London, under the name of the Temple Coffee House Botanic Club. This seems to have come into being as an appendage of the Royal Society, owing its founding to the fact that this august body went into recess for the whole of the summer and thus deprived its botanist Fellows of a forum just at the time of year when they most felt in need of one. Among those known to have belonged – out of allegedly as many as forty in the spring of 1691 – were virtually all of the Capital's then leading botanical figures. Though the interests of the members probably lay less in searching for plants in the field than in growing them in their gardens, they are known to have organised a programme of outings to noteworthy localities in and around the metropolis.[1] Some years later, in 1721, a short-lived Botanical Society was formed among the medical students in London, at the instance of a future Professor of Botany at Cambridge, John Martyn. This had weekly meetings at which papers were read: lists of plants observed on forays into the countryside round about or on visits to distant homes (whence came one set of first records for Lincolnshire, for example) or – from Martyn himself – a "Course of Lectures, upon the technical words of the sciences". Miraculously, one of its minute-books has survived and, partly thanks to this, it has proved possible to identify all but one of the twenty-three known members.[2]

After a lengthy hiatus lasting all through the middle years of that century, from around 1725 up to 1760 at least, a period during which learned activity as a whole notoriously languished in Britain, the founding of the Linnean Society in 1788 again provided London's botanists with a meeting-ground and, for the first time, a journal too.[3] This new body took its name from the

Swedish naturalist whose *Systema Naturae* (10th edition, 1758) and *Species Plantarum* (1753) are universally accepted as the starting points for zoological and botanical nomenclature respectively. It was a botanical society only in part, however, and it took the whole of the world as its province: the flora of the British Isles was never to amount to more than one of its minor concerns (even though J. E. Smith its founder and long-time President became the author of the standard text on the subject). Field excursions, too, formed no part of its official activities.[4] Moreover, like all the other learned societies in the metropolis with pretensions to some standing, it was designedly exclusive. Its relatively high subscription served as much to deter the unaffluent as to meet the cost of supporting it in the lavish state to which it early became accustomed.

The Linnean Society was exclusive in one further important way. Owing its very being to the acquisition by Smith of the collections and library of Linnaeus, it saw itself as the guardian not just of the binomial nomenclature which these uniquely authenticated, but of the entire Linnaean classificatory tradition as well. For some of the members this had become almost a second faith, and one of the more fanatical of them, George Shaw of the British Museum, is said to have gone even so far as to propose putting his heel on any shell not described in the twelfth edition of the *Systema*, proclaiming that "things not in Linnaeus ought not to exist".[5] As a result the Society fell into the trap of championing the old-style Sexual System of classification (based on counting the floral organs), despite its admitted artificiality, and opposing the rising tide of opinion that favoured a classification that sought to reflect the apparent affinities between organisms observable in nature. The first work to be published in the English language that presented an arrangement of British plants by this Natural Method appeared in 1821, under the nominal authorship of Samuel Frederick Gray, a druggist and surgeon in Wapping.[6] In reality, as it soon came to be known, most of this was the work of his 21-year-old son, John Edward. When in the very next year the latter's name was put forward for election to the Linnean, the Fellows of that body repaid him for his effrontery with a humiliating blackballing.[7] The consequences of this were to prove far-reaching. Gray never forgave the Linnean, pointedly refraining from allowing his name to go forward again, even after rising to become Keeper of the Natural History Department of the British Museum. Many of his colleagues there loyally took his side and helped to underline the lasting embarrassment to the Linnean of this estrangement between the country's two foremost institutions devoted to the furtherance of the subject as a whole. At the same time Gray became the hero of the scientific radicals, who found in the Natural Method an ideal progressive cause, symbolising as it did the replacement of artificiality with naturalness and the toppling from their thrones of over-conservative elders. If ever a further body were to be founded that identified itself with this competing approach, then Gray was the person above all others whom its members were likely to want to see at its head.

As it happened, there *was* room for another botanical society in London by the time the Victorian era was about to begin. Several decades of prosperity on the back of the surging industrialism had greatly increased the number of those who aspired to such pursuits; but a high proportion of these, lacking the requisite level of knowledge – let alone the social standing or financial means – were inevitably shut out of the established bodies. As a result a new lower layer of societies was gradually being called into existence. Catering on the whole for the less well-educated and the less well-off, these tended to lay emphasis on self-improvement and attached more importance to maintaining low subscriptions than to having premises, accumulating possessions and employing staff. One such body, earlier than most, was the British Mineralogical Society; founded in 1799, this had an atmosphere of social idealism with strong religious undertones that was increasingly to become typical of the genre. Another, the Entomological Club, founded in 1826, so prized its cosy informality that it purposely kept itself minute and dispensed even with subscriptions. A third, which arrived on the scene later still, in the summer of 1836, was the Botanical Society of London. Even without the stimulus of that undercurrent, the Capital could have expected to have acquired around then a society that concerned itself with botany exclusively. For it was not only the audience for learning that was increasingly fragmenting: science itself was fragmenting into specialist disciplines. Reflecting this, as early as 1823, a Meteorological Society had emerged; three years after that the Zoological Society was born: then, after four more years, in quick succession the Geographical, the Entomological and the Statistical arrived to join them. Launching a learned society, it almost seemed, had become the latest metropolitan fashion.

These developments deprived the Linnean Society of much of its previous following among zoologists, so that it might have decided to confine itself to botany alone thenceforward. Tradition, however, was too strong; and in any case the Society was deeply committed by then to purchasing the Linnaean collections from its founder's executors. The consequent financial strain, coinciding unluckily with the rule of an unimaginative gerontocracy, left it in no frame of mind for any bold rethinking. The field was thus left clear for a specialist botanical competitor.

Botany, it so happened, was enjoying a marked wave of popularity just about that time. The main impulse behind this was the peculiarly lively teaching of the subject by the professors of botany at the universities: W. J. Hooker at Glasgow (later to become Director of Kew), R. Graham at Edinburgh (who had been Hooker's predecessor at Glasgow) and J. S. Henslow, Charles Darwin's teacher and friend at Cambridge. These three had rediscovered the effectiveness of instruction in the field, of the kind that had been practised by apothecaries and surgeons for centuries for training their apprentices in the recognition in the wild of the medicinal herbs. Behind this upsurge in the universities lay in turn a legal accident, an unintended outcome of the Apothecaries Act of 1815. This had the effect of requiring anyone wishing to practice medicine anywhere in England and Wales to gain a qualification, the LSA (Licentiate of the Society of Apothecaries), that was renowned for the rigorous standard it insisted upon in the matter of plant identification.[8] The consequence was that wherever medicine was taught special courses had to be added, and special teachers secured, to provide the necessary grounding in botany. If there was no member of staff willing or able to take this on, a freelance had to be engaged for the purpose. In this way a sizeable body of semi-professionals had suddenly emerged in what had hitherto been more or less exclusively an amateur field. At the same time an entire medical generation found itself exposed to a subject which it would otherwise have had little expectation of encountering, and in the process the interest of not a few was captured and some lasting converts won. For much of the rest of the century, as a result, the ranks of British botany were to have a strongly medical component.

It was thanks to this medical impulse, though at one remove, that the founding of a botanical society for London in due course came about. The person ultimately responsible was a Lambeth general practitioner, James Forbes Young. An ardent geologist and antiquarian as well as an enthusiastic gardener and botanist, Young had for one of his neighbours and closest friends, John Thomas Cooper, an apothecary turned chemical manufacturer turned lecturer in chemistry at various London medical schools.[9] Cooper had early abandoned a medical career on finding, like so many others (Charles Darwin for one), that the clinical procedures of the day were more gruesome than he could stomach. This did not deter him, even so, from allowing two of his sons to try where he had failed. The elder of these, Daniel Cooper, was accordingly put under the tutelage of his friend Young, with whose natural history interests the boy had early shown an enthusiastic sympathy. Blossoming forth as a botanist and conchologist, and warmly encouraged by Young, the youthful Cooper compiled a guide to the plants (and, to a lesser extent, the molluscs) to be found in various select localities within thirty miles of London. Entitled *Flora Metropolitana*, this appeared as a slim volume in 1836, when its author was nineteen or twenty. In a splendid flourish of loyalty Young subscribed for no fewer than six copies. It was not a very impressive little work (H. C. Watson, in particular, was to be highly scathing about its inaccuracies)[10] but at least it gave Cooper some botanical standing. More important, in collecting together the various locality lists of which it was composed, he made the acquaintance of a wide range of fellow enthusiasts in the London area, many of whom were to be early recruits to the Botanical Society that he was shortly to take the lead in founding. Several of those credited with lists turn out to have been residents of Lambeth and were doubtless friends or patients of the ever-supportive Young. In due course the future Society was to boast not only Lambeth's leading doctor and chemist, but also the wife of its parish clerk, the daughter of its Member of Parliament and the former headmaster of one of its schools. Thus in essence, it may fairly be claimed, it was Lambeth-conceived, even if not, in the event, Lambeth-born.

That birth necessarily had to be in somewhere more central, for it was not a merely local society that Cooper was now resolved upon starting but one that drew its support much more widely. It is more than possible that what he originally had in mind was an essentially London body, based on his own personal contacts and those of Young and his father. But at some point in his discussions a more ambitious idea must have taken root and it was decided that the scope could with advantage be made national. No letters of Cooper's (or Young's) bearing on these matters are known, so this can only be surmise.[11]

The reason for aiming at so wide a membership was not *folie de grandeur* (though from what we know of Cooper this may well have played a part) but a practical one: among the planned functions of the new society was to be an organised exchange of herbarium specimens. The wider the geographical spread of the members, the more effective in attracting desired material this was likely to prove to be. A scheme of this kind had been in the air in British botanical circles for quite some years. Extending a herbarium by exchanges with other collectors was of course an obvious thing to try to do and had been a widespread practice for a long time already. But discovering who was interested in exchanging was by no means easy and one could never be sure that distant correspondents would fulfil their promise, let alone supply material of the species one really wanted. Here is an example of the kind of letter that better-known botanists were being plagued with on the eve of the Society's founding:

> Although I have not the honor [sic] of being known to you, yet your name so often quoted as one of the first of our botanists induces me to apply to you for the same assistance which you have so liberally applied to others.

> My object is the completion of my Hortus Siccus of British Plants, and my request the supply of any superabundant specimens which may be peculiar to your neighbourhood or North Wales which you have so successfully examined. In return for them I shall be happy to return any which are found about London or the southern counties of which I have many duplicates.[12]

Around 1826 two German botanists, E.-G. Steudel and C. F. Hochstetter, had had the idea of organising a collecting co-operative, on a commercial basis. They set up a society, the Esslinger Reisegesellschaft, and invited those who enrolled in this to subscribe for shares in collecting expeditions which it was intended it should sponsor; in return, the subscribers were to receive a proportionate amount of whatever the expeditions returned with. The operation itself was termed the *Unio Itineraria* (or 'German Travelling Union', as it came to be known in Britain) and all material distributed bore that name on its labels. The enterprise proved a considerable success and over the years greatly enriched herbaria in many parts of Europe. William Pamplin, the leading botanical bookseller, acted as its London agent and secured a number of substantial subscribers in this country.[13]

Although the *Unio Itineraria* was expressly directed to securing specimens from distant countries (thereby appealing especially to institutions), its example can hardly have been lost on those who felt the need for some more organised method of exchanging material between collectors within the British Isles. Sure enough, in October 1829 a letter duly appeared in the newly-founded *Magazine of Natural History* under the heading 'A depot for the exchange of natural history articles', advocating a scheme on similar lines. There was one novel feature envisaged, however: that participants would mark their desiderata in certain standard published works (which would thus in effect serve as catalogues), each item being marked with a number corresponding to its rarity or other circumstance enhancing its value. The author of the letter disclosed his identity only to the extent of his initials and that he was writing from Edinburgh; but the evidence is sufficient to leave no doubt that this was the first irruption on to the stage on which he was to play so long and dominant a role of Hewett Cottrell Watson, at that date a medical student in his middle twenties;[14] and already embarked on the pioneering studies of the distribution of British plants for which he was soon to become well-known.

For a time nothing came of Watson's suggestion, and two years later there was a forlorn note from him in the same journal soliciting exchanges of specimens on his own individual behalf.[15] Then, five years after that, in February of 1836, a group of botany students at Edinburgh University –

among whom Edward Forbes, later the founding father of marine biology in Britain, is said to have been the leading spirit[16] – decided to start a botanical society in which the systematic exchange of specimens was to constitute a central function. Watson had left Edinburgh by then and was not recruited till the autumn of that year, but it seems highly likely that the incorporation of this novel feature for a society was due to him, even if maybe at one remove. Certainly it is significant that in the first year of operation Watson made more use of the exchange facility than almost any other member.

With a five months' start, the Edinburgh society substantially filled the vacuum that the nation's botanists had been aware of and to such effect that its London counterpart was made to appear almost superfluous when it came to announce its own inaugural meeting for that July. The Edinburgh society had the further advantage of being closely associated with a university – and a university, what is more, that was generally acknowledged as the leader in science in Britain at that period. An impressively high proportion of the upper levels of the medical profession throughout the British Isles had taken their degrees there and provided immediately a rich recruiting-ground for the new society. Even better, the Edinburgh botanists had built up very good contacts with their opposite numbers in other universities, most notably at Cambridge with Henslow and with C. C. Babington, the man who was later to become, by general consent, the foremost authority on the British flora. This connection was strengthened in Babington's case by his absorption into a secret, quasi-masonic society started by Forbes and several of his student friends;[17] his special bond with Edinburgh was to prove unshakeable and was to be strengthened even further when in 1842 he became joint botanical editor of the Edinburgh-identified *Annals and Magazine of Natural History*. Watson was later to become deeply suspicious of this Cambridge–Edinburgh axis, accusing it of being "a small band of active men . . . determined to monopolize the press, in all matters connected with British botany, and exclude all works which do not take up their species and names". "Though few," he insisted, "they have influence by commanding, directly or indirectly, the *Annals & Phytologist*, the Edinb. Socy., and the botl. department of the British Assocn. – They will also determine the books to be printed by the Ray Club which will be equivalent to putting their mark of condemnation against others."[18] He was to attempt at least once to shift Babington from favouring the Edinburgh society exclusively,[19] but the furthest Babington was ever prepared to go was to donate to the London society one year a few of his duplicates.

Its strength in the professional circles that principally counted gave the Edinburgh society an appearance of weight that the London one had no hope of matching.[20] This impression became the stronger when shortly it began bringing out a regular volume of *Transactions*, the material in which was to prove consistently above-average in quality. Indeed, in so far as comparisons were to be made, they should have been with the Linnean Society rather than with the Botanical Society of London. Even so by its mere existence the Botanical Society of Edinburgh undoubtedly did injury to its London namesake, depriving it of numerous members whose involvement would assuredly have bolstered it (for subscribing to both[21] was to be expected only of the keenest) and showing it up as amateurishly second-rate – at any rate initially.

CHAPTER 2

A Society is Born

It is not at all easy to decide precisely when in 1836 the Botanical Society of London should be regarded as having been founded. The meeting advertised for July 27th was to prove only the first of several in which the aims and constitution were debated and progressively hammered out. In subsequent years the meeting that took place on November 29th, four months later, was celebrated by the Society as its anniversary (with a special commemorative dinner);[1] but this would seem to have been contrived, the date having been settled on merely because it was symbolically appropriate, being the anniversary of the birth in 1627 of John Ray, regarded then as he still is today as the father of English natural history.

The venue chosen for this long string of inaugural meetings was the Crown and Anchor Tavern, which then stood on the south side of the Strand, facing St Martin's-in-the-Fields. This was by no means the hole-in-corner beginning that it sounds: the Tavern was a well-known and long-established meeting-place, with an assembly room which would hold at least two hundred. On each occasion W. H. White, private secretary to the Duke of Sussex – the President of the Royal Society – was invited to take the chair.[2]

The July meeting opened with a proposal by Daniel Cooper, seconded by his 'shadow', William Chatterley,[3] that "a Society be formed for the exclusive cultivation of Botanical Science, – to be called the PRACTICAL BOTANISTS' SOCIETY OF LONDON". 'Practical' was one of the vogue words of the period and evidently denoted those who actually engaged in a pursuit – 'got their hands dirty', as it were – as opposed to merely professing it (a Practical Entomological Society was founded in that same year). Thankfully, however, this name failed to find favour, and at the following meeting 'Botanical Society of London' was unanimously adopted instead.

The objects of the Society, it was tentatively agreed in July, were to be "the advancement of Botanical Science in general, but more especially Descriptive and Systematic Botany", by means of (1) the reading of papers on the habitats, particular characters, etc. of plants and (2) the formation of a library, museum and herbarium, for the purposes of reference and the exchange of specimens. This having been settled, a Provisional Committee was appointed to devise a set of rules and draw up a prospectus. Chatterley consented to act for the time being as Honorary Secretary.

On Wednesday, October 12th, a General Meeting took place.[4] This was timed for eight o'clock in the evening, the normal hour for the London societies to forgather at that period. The notice advertising it had been sent out three weeks earlier and had evidently received a particularly wide circulation in the medical world, for, if the report of the proceedings that appeared in the press is to be believed, most of those who attended were from the various London medical schools.[5] The meeting was chiefly given over to approving the draft rules,[6] fixing the levels of subscription, which were set deliberately low (the top rate to be a mere guinea annually after a guinea entrance fee, which was only one-third the cost of belonging to the Linnean), and receiving notice of the donations that were already beginning to come in to the proposed library and herbarium. Seemingly, however, there was little or no discussion of the one bombshell that had been contained in the announcement of the meeting: the proposal "that Ladies be admitted Members, with a full participation in the advantages of the Library, etc."

For women to be admitted to learned societies as members was almost unheard-of at this period. But the founders of the Society were determined that it should not be discriminatory and to allow in women was only carrying this stance to a logical extreme. It was not by any means certain, of course, that women in any numbers would respond, and indeed the response was to prove disappointing all along. Throughout most of the Society's life they probably never comprised more than about six per cent of the subscribing membership, but even as a token innovation it added a certain spice and considerably helped in obtaining publicity.[7]

The meeting that followed, on November 3rd, "a crowded assembly of both ladies and gentlemen",[8] carried matters a stage or two further. It was now agreed to rent a set of rooms at 11 John Street (now John Adam Street), close to the Society of Arts. It was also agreed to meet twice monthly, on alternate Thursdays,[9] from October round to June and once a month during the remainder of the year. At long last, too, the first paper was delivered – appropriately by Cooper. It was on 'The Influence of Light upon the Common Broad Bean' and "excited great interest, more particularly with the ladies".[10] Already, it seems, Cooper was displaying his undeniable flair for teaching. But if this encouraged his audience to suppose that such was the fare that they could expect to be put before them on a regular basis, then they were in for a disappointment; for all the papers at the next few meetings were uncompromisingly in the Society's declared preferred territory of 'descriptive and systematic botany'. Here were the seeds of a detectable undergrowth of disillusionment.[11]

Few, however, can have found of no interest the paper that followed a fortnight later. This was by a Surrey schoolmaster, Alexander Irvine, later to achieve some prominence as editor of the second series of the *Phytologist*. Entitled 'On the Importance of Local Botany', it called on the members to work together to produce a Flora of the environs of London. To this end, Irvine proposed, printed lists should be circulated for people to enter their records on, "such lists having columns for the insertion of the precise habitation of the species, the nature of the soil where it grows, the altitude of its locality, the time of flowering, and such like". It was a pity there was apparently no response to so forward-looking a suggestion.

By now the Society had a President. This was, of course, as it just had to be, the evangelist of the Natural System: J. E. Gray of the British Museum. He was clearly touched to be asked, particularly because, as he was not slow to point out, he was personally unknown to almost all the members – and in any case for a good many years now had been engaged exclusively in zoology.[12] However, as must have been made clear on both sides, he would be functioning only as a figurehead. Despite his extraordinary appetite for work and driving sense of public duty, he simply did not have the time or energy to spare to do more than take the chair at meetings. The Society, in gratitude, did its best to make him feel at home by reserving for an early meeting a paper demonstrating the superiority of the Natural System and by appointing to the initial Council the author of a forthcoming textbook that expounded its principles.[13] In return, Gray brought in as members several of his British Museum colleagues as well as, in due course, his son-in-law and his nephew.

Almost at once, no doubt to his dismay, Gray found himself called upon for a presidential address. November 29th had been chosen as the date of the Annual General Meeting, when the Officers and Council were to be elected and there was the usual need for some special attraction to ensure a respectable attendance. As his topic Gray chose the wide-ranging 'state-of-the-art' survey traditional to such occasions, prefacing his remarks by welcoming the formation of the Society "by a few young intelligent men who felt that there was no institution that exactly suited their views" and pointing out that the number who had already joined was greater than the Entomological Society or even the Zoological had attracted in their comparable inaugural phases.[14] The address was never published other than fragmentarily; the original script, however, has survived[15] and reveals that most of it was rather platitudinous. Probably he dashed it off in a great hurry and was indebted to botanical colleagues for much of the material.

The election of the Council and Officers evidently went off smoothly. Chatterley was confirmed as Secretary, Cooper was appointed Curator (an honorary office during this opening period), while as Treasurer the services were secured – for what was to turn out to be the entire life of the Society – of a Pentonville headmaster, John Reynolds. Those put on to the Council included W. H. White, who had chaired the preliminary meetings, and a West End solicitor, G. E. Dennes, of whom much more was to be heard. The rest consisted of five medical men (T. Bell Salter, J. Freeman, C. Johnson, D. C. Macreight and H. A. Meeson), C. E. Sowerby, a bookseller, a former headmaster Aeneas McIntyre, an actuary J. Mitchell and a freelance naturalist in the person of Edward Charlesworth, who was soon to become well known as the next editor of Loudon's *Magazine of Natural History* but at this time held a temporary appointment at the British Museum.[16] On the pattern of other societies, the choice of the two Vice-Presidents was the President's privilege and as these Gray nominated Macreight and Johnson.

It was a curious and diverse set of people. Sowerby and Mitchell, for a start, seem to have had only the slenderest connection with botany and, mysteriously, left both the Council and the Society within about a year (had they joined under a misapprehension? Or was it some row that occasioned this quick departure?). Of the others, hardly any had any standing as field botanists or indeed as botanists of any kind at all. Apart from the three drawn from the London medical schools, which would be expected to provide a strong contingent, the representation was from other sciences. Charlesworth was almost exclusively a geologist, Chatterley at least equally a chemist and, even more remarkable, no fewer than four, Reynolds, Mitchell, McIntyre and White, were officers of the Meteorological Society of London, a body which was the primary affiliation of at least two other of the earliest members.[17] The Meteorological Society was just at that time being rescued from prolonged dormancy by this team grouped around White,[18] and it is likely that there had been much beneficial interchange between it and the Botanical Society's founders. Special importance was probably therefore attached to giving some formal expression to this particularly strong informal link. The two societies were to remain close for some considerable time to come, sharing the same printer and, in 1840–42, even the same set of rooms (when the Botanists evidently gave shelter to the temporarily debt-harassed Meteorologists).[18]

The weak botanical credentials of this inaugural Council were hardly guaranteed to do the new Society much good in the eyes of the scientific world. Gray, though an obvious choice as President, was in one way an unfortunate one; for an active botanist of stature might have been no less flattered to be asked and would have been far better placed to draw in people of solid reputation in the subject. As it was, the founders' attempts to land some of the big botanical fish, for instance, Sir William Hooker, Professor of Botany at Glasgow,[19] were largely doomed to failure. The only ones who were prepared to join either viewed the Society merely as a potentially useful source of specimens, like Watson, or were naturally sociable figures happy to help it out in a genial spirit of patronage, like the future, long-serving Vice-President, John Miers, who was by profession an engineer. To any who insisted on applying an unindulgent yardstick it appeared (in Watson's later words) "an association of a few tyros", with before it "only an abortive effort and short existence".[20] Several leading London amateurs, let alone those associated with the Edinburgh society or the professionals in the universities, kept well clear of it in consequence. An example, more conspicuous than most, was the philosopher John Stuart Mill, despite the fact that he was busy forming a British herbarium and was a regular field-companion of at least one of the Surrey members, and despite the fact that his half-brother G. G. Mill, probably did join some years later. It was sadly symptomatic, too, that several of the medical members, despite the advantage to their professional image of appearing to embrace as wide a range of scientific interests as possible, omitted to mention their membership when listing other societies to which they belonged in the entries they furnished annually to the *Medical Directory*.

Nor was it just the lack of credibility that caused so many to stand aloof. Cooper's pretentiousness irked the more knowledgeable, especially when, just over a year later, he took it upon himself to give a course of lectures, just before the meetings, on the elements of botany.[21] "He was certainly an intelligent and active-minded young man," Watson was to write of him, patronisingly, long afterwards; "but he had unfortunately imagined himself already competent to guide and instruct others before he had sufficiently acquired knowledge as an observer."[20] In private Watson was considerably more scathing, dismissing him as a "poor egotist" who was "firmly convinced that *he* had given a most unprecedented impulse to botanical science".[22] And if Cooper displayed these failings, it is all too likely that Chatterley was influenced by his example, for with parents who were both celebrated actors he must surely have partaken to some extent of his strongly histrionic background.[23] To this posturing W. H. White, in turn, contributed a further dubious ingredient, with his astro-meteorological convictions and his enthusiasm for a crankish body called the Uranian Society. And as if that were not enough, there was the lingering memory of Aeneas McIntyre's colourful bankruptcy, in the course of which he had sought to evade the Sheriff's Officer by hiding in his garden,[24] to add to the faint air of raffishness.

Yet a further cause for disdain was a wild over-ambitiousness. The Society gave in to the temptation to declare objectives that it hardly looked likely ever to realise, or at any rate with acceptable efficiency, fancying that it could break into an exhilarating gallop before it had mastered even so much as trotting. Indeed, it was apparent at the very outset that the executive officers were being expected to take on far too much, a goodly proportion of which was almost certainly beyond their abilities.

The most extreme example of this overbidding occurred at that first Annual General Meeting. According to *The Times*, "the members expressed a wish for the establishment of a botanical garden on a scale suitable to the Metropolis, and measures are forthwith to be considered for effecting this desirable object".[25] It sounds as if there was an unexpected motion from the floor, perhaps from the medical school teachers, who would have found an alternative to the Apothecaries' long-established garden at Chelsea more than useful. If the proposal did arise in this manner it seems that the Council did not have the strength of mind to ask that it be deferred. In their inexperience they may have attached more value to having a further, luscious membership bait than to maintaining the Society's credibility. Fortunately, however, no more seems to have been heard of the proposal for another two years. When the President then came to broach it, in his 1838 Address, it was to give it a tactful burial:

> With regard to the Botanic Garden, as proposed as one of the objects of this Society, your Council have been delayed from entering into any speculations of the kind until they feel assured of its success, but have determined to proceed surely but slowly in working out that desideratum and have set aside all idea at present of carrying it into effect.[26]

What had rescued the Council, as Gray hinted obliquely, was the entry on to the scene in the meantime of another, much more weighty body which had the establishment of a major botanic garden in London as its sole purpose. This, most regrettably, had decided to call itself the Royal Botanic Society, oblivious of the confusion that so similar a name was bound to cause. It is by no means inconceivable that some people joined the one in mistake for the other, an error made all the easier by the fact that the Royal Botanic Society also arranged lectures and built up a sizeable herbarium.[27] After vainly putting in a take-over offer for Kew, which had fallen on evil days and was still awaiting its salvation by the Hookers, the new body settled for a site of its own in Regent's Park and quickly built this up into a successful and even fashionable enterprise, which was to endure for almost a century.[28] It thereby put paid for good to any lingering hopes the Botanical Society may have nursed of branching out in the same direction – or, for that matter, of adding 'Royal' to its title.

Another instance of over-sanguineness was the scheme for a large-scale exchange of specimens. Success in this sphere mattered far more to the Society than the acquiring of a botanic garden, for it had laid much stress on it as one of the attractions of membership. Unfortunately the Society had not thought out the procedures sufficiently and went on to compound this weakness by entrusting the operation to Cooper, in his capacity as Curator, who was ill-suited both to the hard administrative grind and to the unrelaxing vigilance for which an undertaking of this sort was bound to call. Apart from that, he did not possess the requisite taxonomic knowledge to be able to spot erroneous determinations or confusions in nomenclature. His worst sin, however, was to institute the practice of sending out quantities of the blank labels bearing the Society's printed name to any members signifying their intention of contributing specimens. This opened up a whole Pandora's box of doubt, for not only was there nothing to stop the members concerned from using the labels for non-Society purposes, but it also conferred a spurious respectability on (as Watson was later to put it) "the naming of even the veriest tyros".[20] Subsequent Curators tried to stop the rot by initialling labels as an indication that they had passed through the Society's hands and that the naming was thus authentic; but that even this was far from foolproof is shown by a specimen of Wood Club-rush, *Scirpus sylvaticus*, in Watson's herbarium which had gaily been endorsed as Greater Wood-rush, *Luzula sylvatica*.[29]

But to dwell for long on these deficiencies would risk giving the impression that the Society was a disaster and its founding a regrettable mistake. This was not at all the case. Already, it was well on the way to realising its aim of 'enfranchising' the botanical under-class, by making available to one and all the same range of facilities that the grander societies had long reserved for their members. Admittedly, the library that was fast beginning to take shape was unlikely to possess much that was rare and expensive, dependent as it was so largely on donations. Nor, for the same reason, could all that much be expected of its museum, in the absence of the resources necessary to build up a collection systematically. Even so more was being achieved in these directions than might have been expected, thanks to some handsome windfalls. Probably the finest of these was that from Dr Macreight, one of the Vice-Presidents, who on leaving to live abroad in the summer of 1840 made the Society a present of his choice collection of seeds and timbers and of a very large number of books.

Predictably, it was herbaria that proved the favourite medium for individual generosity. There were, after all, far fewer outlets available to anyone wishing to dispose of one of these and, unless it was especially extensive or possessed some other quality that raised it above the ordinary, it had next to no value commercially. Collections made in distant countries by botanically inexperienced travellers, in particular, feature suspiciously often among the reported donations: those who specialised in British plants (who were probably in the majority) would have been only too relieved to pass these on. Typical was a herbarium made in North America by officers of the Hudson's Bay Company, presented in the course of the first year by J. Freeman. Probably of rather more interest was a large collection of French plants, "supposed to have been part of the herbarium of Jean-Jacques Rousseau", donated in February 1838 by one of the sons of the celebrated American bookdealer, Obadiah Rich (who had doubtless been responsible for its acquisition). Four years later Obadiah Rich himself was to be the donor of some American specimens, but as to how numerous or historic these were we are given no details.

Though most donations were of flowering plants and ferns (especially in the case of non-British material), they were not of these exclusively. At the time the Society was founded microscopes were undergoing rapid improvements and at the same time falling in price dramatically, bringing the study of the minuter elements of nature strongly into fashion. The Society had its share of enthusiasts for these less familiar categories of plants, and there were to be several gifts of substantial collections of mosses from various donors and also of fungi from H. O. Stephens and P. B. Ayres.

Particularly valuable additions to the herbarium came from those who read major taxonomic papers before the Society and afterwards presented the specimens they brought to illustrate them. In 1839, for instance, Mrs Margaretta Riley donated an example of every known species of British fern, as a preliminary to the series of authoritative contributions she and her husband were to provide on this group.[30] In the same way the Worcestershire botanist, Edwin Lees, was to deposit numerous brambles after a marathon stint in 1844 expounding "A Synoptical View of the British Fruticose *Rubi*", the product of his important pioneer researches on this group in the West Midlands. Astonishingly, the greater part of no fewer than seven successive meetings was given over to this. It is hard to see any present-day audience of botanists putting up with such a fare even on a topic more generally in favour.[31]

Papers reporting original work of such a high standard, however, were all-too-rare treats throughout the Society's history and particularly during this initial period, for there were simply not enough members of the calibre to produce them. Indeed, the flow of original material on offer of any kind can hardly ever have been all that vigorous, considering the high proportion of the membership which was too diffident even to take part in discussion and saw itself as coming to be taught, not to do the teaching. To fill the gap, therefore, the Society fell back, very sensibly, on the reading of translations of foreign work. As these inevitably had to be solicited, at any rate at first, this had the added advantage of enabling it to introduce an element of direction into the choice of topics to be put before the members. In this way it proved possible to give the Natural System and its advocates a greater and more emphatic exposure than would otherwise have been the case. Luckily, the Society was well provided with members with a reading knowledge of French and German, if the number who came forward with these translations is anything to go by, though for all one knows some at least may have been ghosted by outsiders. On the other hand, one or two of those contributing them show some evidence of specialising in a particular country's literature (most notably W. H. White, in that of Belgium), which is suggestive of first-hand acquaintance, reflecting perhaps some personal connection.

Unfortunately, after the initial run of meetings there is no mention in the published reports of how many attended (and on this point the one surviving minute-book does not help either), so there is no means of gauging the popularity of different topics. Lectures on far-away places seem always to have been good draws, to judge from the frequency with which Lhotsky was asked to speak on the vegetation of Australia and from the prominence accorded to Robert Schomburgk's account of his discovery in British Guiana of the Giant Water-lily, *Victoria amazonica*. If papers on local botany tended to predominate as time went on, this is more likely to have been because there were many more members able and willing to produce these than because they were the ones in greater demand.[32] We must not in any case underestimate the power of the sheer general enthusiasm evoked in the early stages of the Society's existence to render its audiences indiscriminately appreciative.

If the meetings were indeed as popular as we are given to believe, it is slightly surprising that the Society, considering the kind of body that it was, did not launch forthwith into a programme of field excursions as well. The peculiar delights of these had long been known to the London botanists through the long-standing activities in this direction of the Society of Apothecaries and its various freelance imitators; moreover, the Linnean Society's informal Club had already for some years been setting an institutional example. True, it was only as recently as 1831 that the concept of a field club had been evolved, up in distant Berwickshire, and it was not to be till the middle of the next decade that this model would become fashionable.

Yet the Society had a commitment to informality, liked to see itself as refreshingly innovative and, above all, catered primarily for a membership frustratedly bottled up in the now sprawling metropolis. It was a good opportunity missed.

A modest approach towards something of this kind did begin in 1838, however. Early that summer a 'First Excursion' was announced, under the leadership of the ever-energetic Cooper. The destination selected was "a spot ... called Woking", because of its convenience of access by the new London and Southampton Railway. All seems to have passed off successfully, to judge from the account that was published shortly afterwards.[33] But if 'First Excursion' was intended to imply that this was the harbinger of several more, as most members must surely have thought, hopes had been raised in vain: for there was no follow-up that season. Another 'First Excursion', however, took place the next year, this time to Box Hill and the neighbourhood of Mickleham, and there are mentions of further ones till 1842 if not later. The term evidently denoted a special, ritual occasion – a sort of communal celebration of the onset of the main summer season – and is not to be taken as indicating any more thoroughgoing activity in this sphere.

By contrast, another development around the same time was prosecuted more wholeheartedly. This was the setting up of a network of Local Secretaries, after the example of the Edinburgh society, which in turn must have copied the idea from the geologists. Eight of these had been appointed by the first Anniversary Meeting and their number was to grow to 46 (including several overseas) by the time of the 1843 edition of the Prospectus. By a lucky chance, one of the letters of appointment has come down to us.[34] The duties, it transpires, were no more demanding than to act as a funnel for donated items and subscriptions (this was still three years before the advent of the Penny Post, so collecting these from provincial members was expensive and cumbersome), but there was a broad hint that the occupant of the office was expected to be especially assiduous in recruiting locally.

It is probably no coincidence that provincial membership rose markedly in the year following this initiative. For that twelve-month period the newly created categories of Honorary and Foreign Members complicate the analysis of the figures but, from a tally of the individuals known to have joined then, it transpires that in 1837 recruits had been drawn from the London area and the provinces about equally. In 1838, however, London recruitment slowed and there was a strong expansion in the provinces. Up till then the Society had been predominantly a London one; provincial (or 'corresponding') members – and you could qualify as one of these even if you lived as near in as Reigate or Uxbridge – paid only half as much as 'resident' ones, in recognition of the fact that they were too distant to avail themselves of much that the Society had to offer; and so this big extension geographically was of value scientifically much more than it was financially. Even so the extra income came in handy, enabling the Society to return into the black after a first year in which it had ended up with a deficit of almost £36 (a worryingly large sum when translated into today's values).

On the strength of this heightened prosperity a move was made to larger premises. Some difficulty was evidently experienced in finding a set of suitable rooms in the desirable central situation at a rent that could be afforded. After a shortish stay at 75 Newman Street (off Oxford Street) and an even briefer one at 25 Bartlett's Buildings in Holborn, eventually, around 1840, it found what it was looking for at 20 Bedford Street, in the neighbourhood of Covent Garden, next door to the premises now occupied by Messrs Moss Bros.[35] As a scientific address this had reputable antecedents, since the Geological Society had been quartered there at one time, and it afforded the chance of some useful extra revenue from the hiring out of the meeting-room. As the Society was never to move again, it may be taken that this final home proved satisfactory.

By this time, touchingly anxious for the Society to be considered on a par with the metropolitan societies of standing, the members were showing increasing signs of a concern with indicators of status. Repugnant though this must have been to the puritan instincts of so many of them, the desire to have the Society treated with respect was insistent. An early, over-hopeful step on the part of a few had been to style themselves 'MBSL'; exposure to the test of a public airing, however, had soon put paid to that little vanity,[36] though there were to remain to the end a determined one or two who proudly put 'Member of the Botanical Society of London' beneath their names on title-pages or in communications to journals. Following that the Society even dared to dip its toes in ceremonial: in May 1840, for example, a solemn address was moved congratulating the Queen and Prince Albert on the occasion of their marriage.

But these indicators are trivial compared with what was to come next, in the fourth summer of its existence: the bringing out of a volume of its own printed *Proceedings*. The rates of subscription were far too low, of course, to permit this to be distributed free to members; but fortunately it was quite normal for learned societies at this period to charge extra for their publications, so it can be assumed that all who desired a copy were content to pay up uncomplainingly (as a sweetener they were offered it at a substantial discount, provided they obtained it from the Secretary direct).[37] The volume must have done a lot for the collective ego and it is certainly of great value to the historian (not least for the introductory list of members, the only one ever to have been produced), but it probably proved unsustainably expensive – for at any rate no more were to follow. By good fortune, though, shortly afterwards, in June 1841, Edward Newman launched a commercial monthly, the *Phytologist*, which turned out to serve quite adequately as, in effect, a house journal for the Society, regularly carrying reports of its meetings and honouring with print many a paper that had first been read before its members.[38]

For this slight if definite theatrical trend the personal bent of the two founders (and pivotal figures), Cooper and Chatterley, was doubtless responsible in substantial part. The little that we know of them suggests that by nature they were specialists in public relations ahead of their time. Their influence, even so, was not to outlast this inaugural period; for both, for their separate reasons, were now to disengage and more or less drop out. Of the two, Chatterley's departure was the earlier and more clear-cut: resigning as Secretary at the end of 1837, after only fourteen months in office, he was put on the Council instead – only to resign from that in turn just a year or so later. A paper which he gave in November 1838, on the unlikely topic of the economics of the coniferous timber trade, carries a hint that career pressures may have been the reason. In the case of Cooper, on the other hand, a natural restlessness is more likely to have been to blame. He was clearly the kind of person who finds stimulus in starting things but quickly becomes bored once they have settled into routine – an inventor, not a developer. The first sign that the Society had ceased to hold his full attention is contained in Gray's third presidential address, in which he cites "the numerous occupations of our Curator" as the reason for the foreign section of the herbarium having not yet been arranged. A few months later, during 1839, Gray produced a temporary two-year job for him at the British Museum, putting him to work on the invertebrates; but this paid very poorly, and much of his energy had to be given over to earning what extra he could from lecturing and journalism. Caught up as well in the launching of yet another London scientific society, the Microscopical, a further commitment presently came his way through this: the editing for one of the commercial publishers of a journal of microscopy. By January of 1842, he was sufficiently widely known and sufficient of a professional to be talked of as a possible candidate for the Chair of Botany at King's College London.[39] Instead, however, tiring of the uncertainties of this freelance existence, he decided after all to settle for security, dusted down his medical qualifications and enlisted in the army as a surgeon. In only a short time after that, within a matter of months, the news came that he was dead. The cause is variously given as "a sudden attack of

J. E. Gray, first and only President of the Botanical Society of London.
Oil painting by Margaret Sarah Carpenter,
subscribed for by the members in 1846.

inflammation of the veins" and "a constitutional irritation supervening from a slight injury", diagnoses which one can merely quote in all their antique impenetrability. He was given a military funeral and buried at Quarry Hill Church in Leeds. He was still only twenty-six.[40]

It was a poignant note on which the Society completed the sixth year of its existence. Cooper, with his charm and amiability and his ever-readiness to help with instruction, had been a popular figure – even if there were some who found his airs hard to take. But by then his services to the Society had already been completed. He had inspired its birth and had seen it through its infancy: it was time now for others to take over and guide it a stage or two further onward to maturity.

CHAPTER 3

Watson Takes Over

At the fifth Anniversary Meeting, in November 1840, Macreight retired as one of the two Vice-Presidents, having departed by then for Australia, and Watson was elected in his stead.

Watson had joined the Society probably in the previous year,[1] but he features first in the reports of the meetings only earlier in that very month; so his swift elevation to this honour is testimony of the renown he had won in botanical circles by then, at the age of thirty-six, with the first two in his long series of monographs on the distribution of the flowering plants and ferns of Great Britain.[2] These, the *Outlines of the Geographical Distribution of British Plants*, which he had expanded into a second edition, with a slightly altered title, and *The New Botanist's Guide*, had been the fruits of his Edinburgh student years. He had gone down from university in 1832, without proceeding to a degree – even though he was one of the best students in his class. Ostensibly the reason for this was a visitation of ill-health, but a deeper one may well have been an aversion to following medicine as a career: he had soon discovered, he was later to confess, "that a knowledge of the medical sciences was a very different matter from the practice of the medical art, and that the later would be discordant to my tastes and habits".[3] Fortunately for him his father, an attorney and magistrate in Cheshire, had been comfortably off, and though relations between them had been strained (for the father was a domineering bigot), as the eldest son he had inherited just sufficient means to enable him to live frugally as a self-supported professional scholar – or, viewed from another angle, as an amateur of a kind exceptionally thoroughgoing and dedicated in commitment. On two or three occasions Watson did make overtures with a view to securing a permanent botanical position, but invariably these were half-hearted and it is apparent that he much preferred the slender independence that allowed him to devote all of his time and energies to his research.

Although he had been living not all that far from London, in the north of Surrey at Thames Ditton (designedly close to Kew), for a full three years by the time of the Society's founding, he saw no cause to join it at first, considering it unlikely to survive and in any case superfluous. The Edinburgh one, in his view, offered British botanists sufficient for their needs: if two societies proved unavoidable, that should be treated as the national one and its London counterpart as merely "a sort of local offshoot".[4] About that time, moreover, the then fashionable 'science' of phrenology had displaced botany temporarily at the centre of his concerns. With his classifier's mind Watson found intellectually seductive this system of supposed relationships between various traits of personality and the external conformations of the head; and for long he enjoyed deploying it in his encounters with friends and enemies alike as a kind of all-purpose weapon of analysis, much as a later generation was to do with the theories of Freud. By 1837 he was sufficiently enamoured of the subject to purchase the *Phrenological Journal* and take on the editorship himself. But the unrestrained forthrightness in criticism that was to be his hallmark all through life – and which was to make him so difficult a person to work with – was not long in alienating all too many of its contributors and readers, and after three years he abandoned the increasingly uphill task and returned in some relief to botany.

It was at that point, while he was vulnerably *tabula rasa*, that the Society had the great good fortune to catch him. Very possibly the approach was not his: certainly Gray, for one, would have known of his presence in the vicinity and was well-placed to have heard of his disillusionment with editing. At the same time it is unlikely to be entirely a coincidence that he was beginning to be dissatisfied with the way the Edinburgh society was going. Two out of every three of the specimens he received in return for his 1840 subscription, he was moved to complain to its Corresponding Secretary, were common Swiss alpines, "those most worthless of all foreign plants".[5] A year later he was far from the only one to grumble at a marked falling-off in the quality of its parcels.[6] A year or so after that even the ever-loyal Babington was "beginning to get very angry" with the society,[7] and it was having to apologise in its Report that a temporary breakdown had occurred in its distribution and other services – the result, as it frankly admitted, of incautious overexpansion wholly disproportionate to its resources. Following the dispersal of the original founding group its management had fallen into much feebler hands and it was never to recapture its initial momentum. Consequently "on finding the Edinb. Soc. so uncertain and procrastinating that I could not get the British plants required – that is, the new and dubious things",[8] Watson turned instead to its London counterpart, albeit far from hopefully.

Uncomfortably aware that all was far from well in their own comparable department too, the officers of the London Society lost no time in inviting him to cast his eye over the duplicates then awaiting distribution (which took place each year during December and January). Though conscious that he was rusty in matters of taxonomy and nomenclature, Watson consented to this readily enough, seeing a chance to extend very considerably the data he was accumulating on what he termed the 'topographical botany' of Britain. The collections of British plants held by the British Museum were at this period still rudimentary and much the same could be said of those at the universities, so the herbaria being built up by the two societies were already incomparably valuable for his purposes. Those of the London one, moreover, as he must have realised, constituted the more promising quarry inasmuch as the very inexperience of most of its members meant that it was the commoner species that they predominantly sent in, and these were precisely the ones on which only vague and sweeping statements about the extent of their occurrence tended to appear in the literature. As he explained in a letter to Sir William Hooker, the work would be of assistance to him in his "two favourite departments of botany, namely, the study of localities and geographic ranges, and also the fancy of ascertaining to what extent species may vary in their habit and character". To these ends "the inspection of numerous specimens of the same species, gathered in different places, affords much facility".[9]

The task that he found confronting him, however, proved to be horrific. In their keenness to render the Society more substantial, the members had vied with one another in the sheer quantity of specimens they could send in.[10] During 1839, in particular, the number received as a result had shot up dramatically, to just a few short of 25,000, completely overwhelming the already-struggling Cooper. Even with the Secretary to help him, the Distribution that year had had to be spread over into February.[11] With Cooper's ever-growing commitments elsewhere the backlog had then grown steadily worse, to the point where the Society, in desperation, had been driven to invent two 'Assistants to the Curator' (in the persons of S. P. Woodward and T. Sansom) as a tactful way of instilling more action. But there had not yet been time for these to start work. The consequence was that Watson found himself having to examine and, where necessary, redetermine material representing probably 1,000 species and between five and ten times that number of individual specimens.[12]

To his further dismay upwards of 80 gatherings turned out to have been misdetermined.[13] On top of that there was reason to doubt the correctness of the locality details on many of the labels. Some members, he discovered, sent in ones that were more or less illegible and easily misread; others had fallen into the elementary trap of mingling between the same set of papers unmounted specimens of a single species collected at different times and places. So extensive was the confusion, indeed, that he was later to advise against taking on trust any of the Society's earlier labels.[14]

It was an exasperating struggle, he was subsequently to recall, to counter the lack of method: "the want of order and precision in the Bedford Street management was too usually carried to the lengths of downright disorderliness and slovenly confusion; while the state of dust and dirt in which the rooms were suffered to remain rendered it quite disagreeable to do anything in them". He would have reverted to the Edinburgh society, he insisted, had there appeared any reasonable likelihood of that body mending its ways.[14]

Even so the London society had three undoubted attractions for him. First and most obviously, its rooms were tolerably close at hand, so that he was in a position to influence it in person. Secondly, he found its ethos sympathetic: "I saw among the resident members," he was later to explain, "a spirit of integrity and good-will towards others." This, he felt, gave it a moral superiority on which "I relied for its ultimate success".[14] Thirdly, and perhaps most crucially, the Society was only too ready to defer to him. Unlike the Edinburgh one, it contained no one among the active members who could rival him in authority. This was most important, for despite his liberal sympathies Watson was at heart an absolutist, despite his love of intellectual sparring the wielder of an unbending certainty that could merge into self-righteousness. Mostly this was the product of a rationalistic positivism, but in some degree a streak of lordliness contributed as well – for it was one of the paradoxes of the man that at one moment he would be fiercely assailing established beliefs and at the next moment hitting his fellow-subversives over the head with his aristocratic connections. Over-ready to take offence – maybe out of an inborn pugnaciousness, maybe out of hypersensitivity – he needed a certain distance in his social relationships. Consequently he was no team man and quite unsuited to the compromising world of committees. His preferred habitat, rather, was the margins: when, through some accident of circumstances, he strayed into the centre of things, as he was now to do in the Botanical Society of London, he did so on the basis that this was a temporary incursion only and half-consciously prepared all the while for his withdrawal. For this reason the only office that the Society could ever persuade him to accept was the largely decorative one of Vice-President. He did not want to be seen to be officially involved and committed: he was merely the visiting consultant. His reluctance to take paid employment was all of a piece with this, reinforced in turn by an underlying austerity. Watson positively preferred his life of barely gentlemanly poverty, just as he was content with barely sufficient print-runs of his books. It was alleged that he made his works scarce quite deliberately;[15] the truth, more probably, was that he was simply one of nature's Roundheads, who thought only in terms of bare essentials.

By singular good luck the 20-year-old who had succeeded Chatterley as Secretary, and was thus the Officer to have to work with Watson most intimately, was perfectly suited to the role. George Edgar Dennes was meek, biddable, unassuming, conscientious to a fault: the natural task-master had inherited the ideal slave.

All references to Dennes agree in testifying to his extraordinary devotion to the Society and its officers. Even the hard-to-please Watson found little but praise for him, at any rate in print: "the ever-working Secretary", "earnest and indefatigable", "of obliging disposition", whose "disinterested exertions have conduced so much to the Society's efficiency". Indeed, the time and energy he gave to his Secretarial duties suggest that his heart can hardly have been in his profession of solicitor; maybe it had not been of his choosing and maybe it was a family practice in which he was indulgently underworked, though, even if so, he was still "much tied by professional duties", according to Watson.[16] More certainly, he could not have given so much of himself to the Society had he not been, like Watson, a bachelor. Like Watson, too, he seemingly derived a certain relish from swimming against the social current, the pair of them regularly choosing the morning of Christmas Day as the time to go through the preliminaries to the annual Distribution.[16] But perhaps his real saving asset, ironically, was that his interest in botany as a science was only slight. His field experience seems to have been minimal and, according to Watson, he could never name a plant from a book description.[16] On Watson's knowledge in these areas he was consequently all-dependent, while making himself no less indispensable to Watson in return by acting as his ever-attentive clerk of works. The two of them together made an unbeatable combination.

Having placed the Society almost embarrassingly in his debt by his Herculean labours, Watson was at once in a strong position to start insisting on improvements. The most urgently necessary of these, he soon persuaded the Council, was the recruiting of a paid Curator. The degree of drudgery entailed in the annual Distributions as well as in the day-to-day maintenance of the collections was far greater than could reasonable be expected of anyone working voluntarily. There was, however, the problem that as long as the Society adhered to its principle of keeping the subscription low the amount it could afford to pay someone, namely £30 a year, was sufficient to buy the services for three days a week of only the most frugal among qualified persons. Explaining this, Watson wrote during that February of 1841 to J. H. Balfour (who succeeded Sir William Hooker as Professor of Botany at Glasgow in that year) to see if he knew of anyone in Scotland who might be suitable, adding that ability to write French and German was essential. Significantly, he was identifying with the Society already to the extent of referring to it as 'we'.[17]

By April the seeming paragon had been found and was soon in position. His name was Joseph Geiger, but apart from that nothing else about him is known. He may well have been a refugee liberal from Germany, desperate for a job; at any rate he must have had the stipulated languages, for 'Curator and Foreign Secretary' appears as his title in that year's Prospectus. Alas, though, he did not last very long; and within a few months the Society was on the search yet again.

The next person to be appointed, apparently later in that same year, 1841, was much more of a known quantity. This was Arthur Henfrey, a recently-qualified surgeon who because of severe asthma found it impossible to practise. Well-connected and with some small private means,[18] he was able to afford the risk of constructing a substitute career out of a plurality of part-time posts with a certain amount of writing and teaching on top. Genial and obliging, he was a more than competent botanist, albeit his botanical leanings were in non-Watsonian directions. He started well and energetically, collecting interesting plants from the London area for the Society's herbarium, some of which, like the Loddon Lily, *Leucojum aestivum*, from Greenwich marshes, he exhibited at the meetings; he also bid actively on the Society's behalf at the auction in April 1842 of the great library and herbarium of A. B. Lambert, who had been Vice-President of the Linnean Society for nearly fifty years.[19] He was just too good, obviously, to last. And sure enough, sometime in the next two years, his resignation took place. But though he went on to higher things, he was luckily not lost to the Society entirely, for he continued to serve on the Council and later as a Vice-President too, contributing several papers to the meetings. After editing the sadly short-lived *Botanical Gazette* and producing several textbooks, he eventually succeeded Edward Forbes as Professor of Botany at King's College London. But ill-health continued to dog him, and he was to die, in 1859, at the early age of forty.

The hunt for a replacement accordingly began anew. This time, however, it met with no success. In July 1844, while botanising at Braemar, Watson encountered the up-and-coming young Shetlander, Thomas Edmondston, "with a collecting box almost as large as himself",[20] and impressed by his ability tried to interest him in the job. But though Edmondston was keen to come to London and to try to make his way as a professional, the pay appeared to him too low to manage on.[21] In his place, Hooker suggested another young Scot, Alexander Croall, a village schoolmaster living near Forfar and already one of the Society's Local Secretaries; but Watson doubted if this man's knowledge was adequate and in any case did not think the proposal that he should combine the Curatorship with making a business of buying and selling specimens would prove acceptable to the Society.[21]

The work of the Curator accordingly devolved again on to the shoulders of Watson and the overburdened Dennes. But although Watson never ceased to grumble about this in his letters to his friends, the development was probably not altogether unwelcome to him. Free of the chore of training the successive appointees and of the uncertainty about how long they would last, at least he now knew where he stood. Moreover, in exchange for all his time and trouble he had been able to strike a bargain with the Society whereby he could take from the stock of duplicates anything that he wanted for himself and, in equal measure, for any of his friends or correspondents – a kind of botanical *droit de seigneur*.[22] As a result his own herbarium speedily became much the most widely representative of the British flora of any in public or private hands (though in taxonomic interest it never came to rival Babington's).

Meanwhile he had been steadily moulding the Society to his design in a variety of other ways. In the autumn of 1841 the Council had obliged him by decreeing that in future as each parcel of specimens sent in came forward for examination the labels were to be written there and then and physically attached to them. From this a great improvement in accuracy resulted, Watson claimed.[23] Around the same time contributors were required to enclose two labels for every specimen, one of these for preserving in a new Society 'label-book' which was envisaged as building up over the years into a valuable register in its own right of distributional data.[24] In addition, also expressly at his instance, an eight-page printed circular with the title *Instructions for the Selection and Labelling of Specimens Intended for Presentation to the Herbarium of the Botanical Society of London* was distributed to all members. This now excessively rare publication was full of quintessentially Watsonian features.[25] After an opening section urging potential contributors to extend their collecting to specimens in different stages of development and to well-marked variations from the type, he launched into his favourite theme: the need to make the Society's herbarium a repository of "more complete information respecting the local botany of the British Isles, ... calculated to assist those investigations into the laws which determine the geographical distribution of plants". To this end he asked for examples of both rare and common species to be sent from different localities, with their habitats carefully noted on the labels. The localities were to be made reasonably precise and a note added of height above sea level, soil type and degree of humidity and shelter. In a closing section he proposed that the Society should also seek to put together a special collection of local herbaria, in a series of bound volumes, "as a substitute of the best kind in place of Local Floras" – for "the specimens would be there to prove the accuracy of the lists". Some guidelines on what these should contain then followed, with the assurance that the Society would undertake the arrangement, indexing and binding. The booklet thereupon ended with the following stirring peroration:

> It is trusted that the Contributors of these Local Herbaria will find a recompense for their exertions, in the gratification of learning thoroughly the botanical productions of their own neighbourhoods, and in the consciousness that much of the information so acquired will become (through their contribution to the Society) a permanent addition to the general stock of scientific knowledge, to be transmitted to future generations.

One would like to think that many of the members read these words and were duly inspired. Certainly a collection of local herbaria was soon after embarked upon, and by the time the Prospectus for 1843 came out the Society was able to report that several of these had already been promised, "from tracts of country whose Floras are unwritten". At least three are known to have been delivered.

At the same time Watson took steps to enhance the quality of the Society's herbarium by badgering leading botanists, such as Babington, Balfour and Henslow, to present to it their spare duplicates of plants of particular interest. Experts on the difficult groups, too, were asked to put together sets of named material for the guidance of the more sophisticated of the members, with a view to stimulating more work on these and increasing the areas of the country in which they were being subjected to intensive study. Prime beneficiaries of this were the Roses and the Brambles.[26] Similarly, the discoverers of species new to the British flora or of outstanding rarities in additional stations were written to and cajoled into sending a specimen for exhibit or, far better, a supply for distribution, in blithe disregard, alas, of the damage this might do to the species' continuation. This practice was inaugurated in 1842 with the distribution of the sedge *Carex paupercula* from Northumberland, after its discovery there the previous year. Next season, the Isle of Wight Calamint, *Calamintha sylvatica* subsp. *sylvatica*, was provided by Bromfield from its lately-found colony there, and David Moore posted from Westmeath in Ireland the first of two consignments of another new sedge,

Carex appropinquata. A steady trickle of others then began to follow: in 1844 the Cut-leaved Germander, *Teucrium botrys*, from Box Hill and the Teesdale Sandwort, *Minuartia stricta*; in 1845 yet another sedge novelty, *Carex montana*, from near Tunbridge Wells; in 1848 the Breckland grass *Apera interrupta*. By 1850 the Society had become so recognised as a medium for the publicising of such discoveries that it enjoyed the privilege that year of having the news of the finding in the British Isles of the waterweed *Najas flexilis* – near Roundstone, in the west of Ireland – first publicly disclosed at one of its meetings. Its discoverer, Daniel Oliver (later of Kew), was only twenty at the time but already one of the Society's most active members. "The circulation of specimens through the Botanical Society is truly the very best method for making known new discoveries, and correcting errors of nomenclature," the report of the Herbarium Committee for 1844 confidently proclaimed, with a forthrightness that could have come from only Watson's pen. "By this step," it went on, "the discovery or correction is promptly placed before the eyes of numerous active botanists, in the best possible form – that of actual specimens in proof of its reality or truth. While the unvarying regularity of distribution, which has hitherto so peculiarly and exclusively distinguished the Botanical Society of London" – an obvious dig here at its Edinburgh rival – "affords a strong additional inducement for making this Society the general centre of inter-communication between the botanists of Britain."[27]

Not all the novelties were brought before the Society intentionally. A Skullcap, for example, sent in from Hertfordshire by a novice member, under the impression that it was the common species, was found to be *Scutellaria hastifolia* – though whether it had really been collected where its contributor supposed remains to this day a mystery. Similarly, a Cranberry sent from a bog in North Wales turned out to be the naturalised American species *Vaccinium macrocarpon*. Such were the gains from having someone of Watson's acuteness as overseer of the herbarium.

Foreign plants were not neglected either and especially if they were from areas in which Watson had an interest personally. As a result of a voyage made by him in 1842 to the Azores (the only occasion on which he ever botanised outside Britain) he was able to prevail upon Her Majesty's Vice-Consul there, T. C. Hunt, to join the Society and start dispatching to it sample collections twice-yearly. With high hopes that the flora of these islands would prove exciting in its geographical affinities, Watson went out of his way to encourage him by sending out the necessary apparatus, sets of named specimens and copies of books likely to be of service.[28] Hunt diligently responded, with collections which by 1846 were running into "many thousands" of specimens yearly. In the same way, if rather less elaborately, another of the members, Adam Gerard, was able to organise shipments of plants to the Society from his brother-in-law in Sierra Leone. By these devices the Society had the benefit of its own, exclusive, on-the-spot collectors without the cost, which it could never have afforded, of dispatching and maintaining these.[29] It was thus able to present itself as a kind of *Unio Itineraria* in miniature.

But it was not enough for specimens of greater interest to be distributed: the members needed to be told what it was about these that made them so especially interesting. In addition, any comments that those who received them might have to add – particularly if these involved querying the naming – deserved to be placed on record. Accordingly, in 1845, Watson instituted the practice of publishing each year short notes on a select number of the plants that had been distributed. During the lifetime of the Botanical Society of London these appeared in the *Phytologist*, but in later years they were to become the subject of a separately published report. This had the incidental advantage that it could be readily cut up and the individual notes attached to the particular specimens to which they related in members' herbaria. As it was around these reports that the great build-up in membership after 1900 was to be engineered, this was to prove the most far-reaching of all Watson's innovations. It was also to prove a remarkably enduring one, for the commentaries appeared without interruption (even if not strictly annually) until the final abandonment by the Botanical Society of the British Isles of exchange activity after the Second World War. Through all this time they constituted perhaps the single most fruitful source of new information on the taxonomy of British vascular plants to appear on a regular basis.

Almost equally long-enduring was the *London Catalogue of British Plants*. The production of a standard list of scientific names, on which contributors could mark the species they particularly wished to receive, was one of the first essentials for a body that had the stimulation of exchange as its central purpose. Recognising this almost immediately on its founding, the Edinburgh society had appointed a committee to work towards this end. Of the eventual product some 4,500 copies were printed, and offered for sale not merely to the members but to the botanical public more generally. That catalogue was speedily found invaluable for a whole range of purposes (not least for marking off what occurred in a particular locality or county, as an ancestor of the modern field card) and copies were soon being bought in quantities at a time. John Stuart Mill, for instance, started off a letter to George Bentham in 1843 with the words: "You probably have abundance of the Edinburgh Catalogues..."[30] It accordingly seemed superfluous for the London society to produce a separate version of its own, even supposing that there was not a strong argument in any case for having all British botanists employing an identical set of names. The furthest it therefore went was to issue a printed sheet, drawn up at its request by Cooper, giving the names in current use, arranged under both the Natural and Linnaean Systems, down to the level of genera only. This was intended merely "to facilitate the arrangement of the British species in the Society's and members' herbaria" and was expressly for use in conjunction with the Edinburgh (species) list.

The Edinburgh catalogue, however, had the odd feature that the names were listed alphabetically (as on a modern field card), instead of in systematic order. The more he came to use it, the more irritating Watson found this. And eventually, in the Christmas week of 1843, he was driven to prepare an alternative better suited to his needs, which the London society thereupon published. Apart from being scientifically arranged, this *London Catalogue* was considerably more discriminating in what was admitted as British; it also introduced the practices of distinguishing the putatively native species from the naturalised ones by means of different type-faces and of including the number of local Floras in which each received mention as a measure of comparative frequency. But despite these improvements it only very gradually displaced its Edinburgh rival. As late as 1849 Babington was denouncing it as "bad",[31] disapproving of Watson's resistance to many of the name-changes that he had shown in the literature to be necessary (in the light of his policy of rigorously scrutinising the credentials of British plants claimed as identical with Continental species). New names and, even more, subdivisions (or 'splits') of long-familiar entities were anathema to anyone like Watson struggling with the task of compiling data on geographical distribution, and the frequency with which Babington was responsible for these only served to heighten the antagonism between the two. There were species, there were subspecies and there were "Bab-ies", Watson once complained wittily to Balfour. But in the end it was Watson who triumphed: the Edinburgh catalogue died at last with its fourth edition in 1865, while the London one continued for sixty years longer and achieved no fewer than eleven editions (excluding several separately-dated reprints).

There were obvious dangers for the Society in allowing itself to bear the imprint so markedly of a single individual, and particularly of one so pronounced in his opinions and so ruthlessly unsparing of the feelings of others. In his commendable concern for accuracy Watson was never one to compromise, and he did not shrink from causing offence by refusing to take on trust the unsubstantiated claims of even the most experienced and competent. The appearance of his *magnum opus*, the *Cybele Britannica*, from 1847 onwards brought matters to a head in this respect. Edwin Lees, for one, went so far as to vent his annoyance in print at the extent to which his records had been studiously disregarded,[32] and almost certainly his sudden disappearance after that date from the reports of the Society's activities betokens his resignation – so fully had the Society come to be identified in his mind with the author of the *Cybele*. Alexander Irvine was another member to drop out for seemingly the same reason: his later attacks in the *Phytologist* on the Society and Watson jointly are otherwise inexplicable in their ferocity.

But it was not only Watson's scepticism that was the cause of people falling out with him. His obsessive concern for system made him jumpily impatient with anyone ill-organised or careless, triggering off upbraidings that were often decidedly peremptory.[33] For the same reason he destroyed without compunction any specimens sent in for distribution which were "either supernumerary or unsuitable or else labelled at variance with the 'Regulations'", a copy of which, as he coldly pointed out, had been "placed in the hands of each member, and the better observance of which might reasonably have been expected". Several thousands of specimens were discarded by him each year on these grounds alone.[34] A great many more were consigned to oblivion for no better reason than that they had not been folded or bent so as to fit the size of the Society's storage paper, as prescribed in the 'Regulations' similarly.

In view of the sheer magnitude of the task he had taken on himself, however, perhaps he can be forgiven much of this high-handedness. The organising of the Distributions alone, as he once had cause to remind his friend Sir William Hooker, was very far from being the usual easy procedure: "to receive a batch of specimens from a travelling Collector, divide them into sets, and pass them on to subscribers." Instead, it involved having to take account of "a hundred different sets of desiderata, to be supplied species by species" – a hugely troublesome job by comparison. "Exact knowledge of species," he emphasised, "ready recollection of the place of each in system, methodical management, rapidity of thought and act, are needful to accomplish the work."[35] Apart from that it was essential, in his view, that he should be ever on hand, "to apply the spur or touch the reins".[36] Even the reports put together for the periodicals by Dennes needed to be vetted by him, he had learned from experience.[37]

The Society was far from unappreciative of all these efforts, and tributes to the immense improvements he had wrought in its affairs were beginning to appear in print more and more frequently. But somehow these did not seem quite enough. It was therefore decided to invite members to join in subscribing the cost of having his portrait painted professionally, as a more concrete testimony of their gratitude for what he had done. To avoid seeming to slight the President by singling out Watson for so special an honour, it was tactfully proposed that a portrait of Gray should be commissioned as well. There was a favourable response from sixty-eight members, and at the Anniversary Meeting of 1846 the finished paintings, by Margaret Carpenter, were ceremonially unveiled.[38] Watson and Gray thereupon presented them back to the Society, for hanging on the walls of the Bedford Street meeting-room. There, as the result of a further appeal, a companion portrait of Dennes arrived to join them in the following April – and if Alexander Irvine is to be believed, portraits of the Treasurer, the Curator "and other patrons of the Society" in due course came to hang alongside too.[39] Irvine, who detested Watson, was later to ridicule this as a piece of vain self-display on the part of those depicted;[40] but if vanity there was, it was surely much more that of the members as a whole, with their persistent weakness for emblems of grandeur.

Presenting Watson with his portrait, however, seems to have been quite fatal. For in the very next year there came the first hint that he was beginning to feel that his task was done and it was time for him to withdraw. This took the form of a long, reminiscing account in the *Phytologist*, contrasting the Society's state before he arrived on the scene with the high level of efficiency to which he had succeeded in raising it. As a result it could "now be deemed the grand centre from which the herbaria of British botanists are supplied" and its over 200 members had come to include "many of the most active field collectors and best practical botanists of Britain". In its exchange activities at least the Edinburgh society had been well and truly outdistanced by it. But the paper did not stop at this fanfare of self-congratulation: its more basic purpose was to put over Watson's by now firm conviction that the Society was valueless except for this "highly important and useful" function of distributing British material. In other words, by rescuing this side of its activities he had in effect been the salvation of the Society itself.

He returned to this theme, and more forcefully, in a second lengthy contribution to the *Phytologist* two years later, early in 1849. The Society, he went on to contend, could command neither the funds nor the labour to achieve the *general* herbarium of reference that it had insisted on aiming at. The only result of the unwise attempts to attain this (and other objectives beyond its capability) had been to impede the one thing it was useful for, namely the distribution of British specimens. Because this view of his was not generally shared, but even more because he had not found "a *sufficiently systematic* co-operation with my own efforts", he proposed to disengage henceforward from any further involvement in the Society's management, even it might be from membership itself.⁴¹ Thereby he would not be in any way "answerable for the dead-lock, towards which some other members are fast forcing it". For want of a competent Curator, he very much feared, "the whole machinery and action of the Society" must shortly be arrested.⁴² What was needed instead, he urged, was for 25–50 of the more select to break away altogether and form themselves into "an exchanging club" quite apart, free of the encumbrance of a herbarium, a library and meetings. "*Tempora mutantur*," he concluded: "the objects for which scientific societies used to be instituted are now better effected by periodical literature, by travelling, by correspondence, and by exchanges. Collective libraries are still important; but we *have* one for botany at the Linnean Society, and *cannot* have one at the Botanical Society of London."⁴¹

With uncanny prescience Watson had foretold almost exactly what the future was to bring. But seven more years were to pass yet before the Society arrived at its anticipated doom.

CHAPTER 4

The Other Side

As Watson had been forced to admit, the strength of his argument was lost on at least the majority of the members on the Council. Their idea of what the Society could best be doing was not at all his, and they were far from open to conversion.

One reason for this is not hard to find. Only London residents, it had come to be accepted, were able to attend meetings with the necessary regularity to be, in effect, eligible for the Council.[1] In the Society's early years this had scarcely mattered, for its centre of gravity had been heavily metropolitan. But thanks to the exertions of Watson many more of the abler field botanists all around the country had latterly been garnered in. As a result, from constituting almost half the membership in the inaugural period the proportion of Londoners ten years later had dwindled to less than a third.[2] By that time relatively few of these London members were active collectors or had an interest in receiving or contributing at any rate British specimens: for most of them the Society was rather a place to resort to for some mild self-improvement after dinner, an appendage to Clubland. Whereas members in the provinces thought of it as a distribution service, its image in the metropolis was more that of a superior evening institute.

As few out-of-town members, one must presume, attended the annual meetings in November, any differing view of the Society's proper role had little chance of being ventilated – or rather, to be precise, Watson's voice was liable to be a lone one. Conveniently ignoring the narrowness of the suffrage, the Council could always counter opposition to its policies by hinting that it had been elected democratically and that it was up to those who disagreed with these to alter matters through the ballot-box. Without a referendum of the entire membership by post Watson had no hope at all of loosening the grip of this London caucus – even supposing there was a sufficiency of country members who shared his views more than very half-heartedly. And of that the evidence is non-existent: so far as one can judge from the contemporary correspondence and published literature, most if not all were content to let things be, unconcerned with how their subscriptions were spent, provided only that they continued to receive an annual parcel of worthwhile specimens. In so far as any of them did have fault to find, it was rather with the very success that the Society had achieved in this department. As one of the more piously Evangelical was moved to grumble, it had "made that a business which used to be a private pleasure" and botany was the worse for this, for "people are apt to become selfish where exchange is made".[3]

Although the attitude of the London members was always likely to be different, if only because of their far readier access to the meetings and the collections, this difference had been becoming much sharper. In the Society's earliest years Cooper's personal enthusiasm for fieldwork had ensured that people who shared this taste were plentifully represented. In addition, apart from true field men, there had been an ample leavening of dedicated, out-and-out collectors – such as Revd F. J. Stainforth, who was to move on from botany to shells and from shells in turn to stamps, to become, significantly, the founder, in the early 1860s, of the world's first stamp-collectors' club. At the same time the Society initially attracted a noticeable number of scientific generalists: not just the powerful meteorological contingent mentioned already, but many others who might equally well have joined a society concerned with some quite other branch of science had there only been one available. Some of these were dependent on scientists for their living, as instrument suppliers or specialist illustrators, and doubtless an element of commercial calculation entered into their enrolment; others, however, are likely to have joined from less venal motives, regarding the Society as just another stepping-stone on the path to self-education. Many of the latter may have owed their membership to the mechanics' institutes, the WEA of the day, for by this period these were patronised largely by clerks and shop assistants in place of the artisans for which they had been intended; the Treasurer, John Reynolds, moreover, was one of the pioneers of the evening education movement.[4] More generally, though, there seems to have been in the metropolis a surplus of scientific talent which was not being catered for institutionally. The best indication of this is the remarkable number of the early members who give evidence of a penchant for mathematics, not an obviously relevant attainment for a botanist at that period, nor for a field botanist indeed even down to the present. Reynolds, for example, was the author of textbooks on geometry, arithmetic and algebra; McIntyre made a hobby of abstruse calculations on local taxation; Hopkins published a work entitled *The Cardinal Numbers*; two other London members were actuaries by profession. The same tendency, admittedly, can be found among the country members also – one was subsequently to spend "almost the whole of his spare time in working at differential and integral calculus, conic sections and the like"[5] – as well as in a rather later intake (J. H. Wilson, we are told, graduated from Oxford "with distinguished mathematical honours"), but it was only among the first wave of London recruits that it appears so pronounced. The complex precision of botanical classification, at a time when classification tended to dominate the subject, no doubt held an appeal for such intellects. On the other hand they were also the kind of intellects that in the provinces just then could find full play in the debating-chambers of the literary and philosophical societies. There was no counterpart of these in London, where, because the learned community was larger, more specialised societies had been formed earlier; men like Reynolds, McIntyre and Hopkins had no alternative but to confine themselves within societies of narrower scope, and in these cramped circumstances the fit between type of mind and type of institution was inevitably poorer and less potentially creative.

By the middle of the 1840s most of these scientific generalists had dropped away, perhaps having found the Society's concerns too oppressively floristic. The field botanists of London, too, with Cooper no longer to energise and hold them together, had ceased to be in such evidence. In their place a new dominant set had arisen: a forceful alliance of lawyers, physicians and engineers who by bringing to bear 'useful knowledge' hoped to clean up the social mess inflicted by a swamping industrialism. Science as a whole had been embraced by their ideology and so fashionable a part of it as botany was unlikely to escape, even if its power to be of help was at first sight not all that apparent. In any case the Botanical Society, by virtue of its conspicuously progressive stance, had already attracted to itself many who were especially favourably predisposed. Thanks to the legal accident of the interlinking of botany and medical training it possessed a goodly share of the ameliorist surgeons and physicians; its Secretary, moreover, was a politically-minded lawyer (for Dennes acted as a Parliamentary agent as well as practising as a solicitor) and so well-placed to recruit in the reforming circles of that other spearhead profession. Reformist ideas had permeated the Bar in particular, flooding it with aspirant social engineers, who were using it as an *entrée* to employment by the proliferating government and lobbyist committees, very much on the pattern of present-day Washington. An upsurge of barrister members in a scientific body at this period, such as has been demonstrated to have occurred in the Royal Institution,[6] is consequently the clearest indicator we can have of an influx of the new creed's adherents. In 1845–6 more barristers, six no less, joined the Society than in all the other years of its existence put together, none of them, what is more, with any reputation as botanists previously. Almost at once two went on to the Council, to be joined on it presently by a third. The cousin of one of these, a future public analyst and Medical Officer of Health, was already on. A leading epidemiologist and a sanitary engineer were elected to it shortly afterwards. By 1849 the newcomers were in the majority and potentially assured of control. Thanks to its very liberality and its scientifically undemanding character the Society had always lain open to a takeover by interests essentially foreign to it. Now, without a murmur of protest, at any rate in the written record, it had finally succumbed.

One sign of the resulting shift in policy is the arrival of a pronounced ameliorist tinge to the topics chosen for the meetings. Thus, in 1847–8 the nature of the mysterious affliction that had laid waste the potato crops and brought on the terrible Irish famine was vigorously debated, on more than the one occasion. The next year J. W. Rogers stirred up further excited controversy with a two-part paper advocating the use of 'peat charcoal' for the deodorizing of London's sewage, attracting an audience which extended to many outsiders and inducing even the great public health reformer, Edwin Chadwick, to honour the Society with his presence. The year after that A. H. Hassall, at successive meetings, mounted an *exposé* of the widespread adulteration of the sugar and coffee then sold in the shops. An abstract of the second of his lectures, distributed hopefully to the newspapers by Dennes, proved such a hit that not only did most of them give it space, but some even followed this up with thundering leaders denouncing the scandal.[7] The delight of Dennes at this must have been unbounded: here was his beloved Society performing at last on the wider national stage – and doing so, better still, in the role of champion of the public good.

What Watson thought of this takeover we do not know. Though he can hardly have been unaware of what had happened, his letters are strangely silent on the subject. In theory, as an active, life-long liberal he ought to have welcomed this commandeering of the Society on behalf of the very principles that he stood for. In practice, it was a development that he was bound to see as inimical to his own botanical ends. Not only was the Society pulled right away from the course that he had so diligently been charting for it: the new men who now held sway were of a very different stamp from their predecessors, being confident members of the senior professions who were far less readily overawed by dogmatic bluster and superior knowledge. In any case their approach to botany was so widely divergent from his that in effect they looked right past him. Faced with a foe who eluded all engagement, torn between his public beliefs and his private ambitions, Watson must have found himself in a situation that was increasingly intolerable. Seen from this perspective, that announcement in 1849 of his intention to withdraw from further involvement in the Society's management looks much less surprising.

To add to his discomfiture, much of the Society's *ancien régime* was on the side of the incomers emotionally: indeed, it is unlikely that their preferences could have prevailed so rapidly had there not been an atmosphere sympathetic to them already. A feeling of aimlessness on the part of the London members may to some extent have contributed. There may also have been resentment at the over-free hand that Watson had been given for his crushingly restricted aims. But the fundamental factor was that several of the leading figures had all along been ameliorists too under the skin. Dennes patently was one – though he could hardly parade these sympathies too openly in view of his longstanding ties of loyalty to Watson; even so, he did not exactly conceal that he acted as legal adviser to the Westminster Reform Society. Gray, similarly, had been active since 1843 in the Metropolitan Improvement Society, through which he had come into contact with Edwin Chadwick; long interested in social, educational and sanitary reform, an advocate in particular of public parks and of more generous opening hours for art galleries and museums, he also claimed to have been the first to suggest a uniform rate of letter postage, to be prepaid by means of stamps.[8] Reynolds, too, the third of that enduring triumvirate, was, as already noted, a pioneer of the evening education movement.

Reinforcing these, in the general body of the membership (a list of members is given in Appendix II), were to be found further like-minded people. Thomas Twining, for instance, who was so assiduous an attender of the meetings as to be asked to take the chair on one occasion in the absence of all of the Officers, was a leading promoter of technical instruction for the poor. Technical education was a favourite cause too of James Heywood, who was a proponent as well of free public libraries. T. J. Dyke, an ardent sanitarian, was another of the earliest Medical Officers of Health. Thomas Beesley used his training as a chemist to give his services voluntarily as a public analyst, later becoming chairman of his local Board of Health. T. G. Rylands, whose entomologist brother sat in Parliament as one of the Philosophic Radicals, had taken part in the anti-slavery movement. E. G. Varenne championed the cause of total abstinence and fostered a Band of Hope ... All through the Society one can trace this strand of public-spirited earnestness.

Side by side with it, and extensively interwoven, ran deep religious convictions – especially (though by no means exclusively) on the part of the Nonconformists. There were 44 of these in the Society at least, out of the putative total of 371, a notably high proportion – and in view of the fact that evidence of religious affiliations is exceptionally hard to come by it may understate the true figure by a sizeable margin. Almost half were members of the Society of Friends, that traditional wellspring of ardent naturalists. The rest were distributed among a variety of other sects, of which the Unitarians were perhaps the most numerous.

The twenty-one Friends formed a noteworthy element in the membership for another, quite different reason: as a result of the longstanding custom among Quakers of marrying only among themselves, a very high proportion were related. No fewer than six out of the twenty-one, indeed, belonged to a single great conglomerate of cousinhood centred on the Marriage family of Chelmsford.[9] But although most marked among the Quakers, this was a characteristic that extended through the Society much more generally. Astonishingly for a national body, fifty-one – that is, fifteen per cent of the total number identifiable – turn out to have been linked by blood or by marriage to at least one other member. (And that figure, it must be stressed, is but a minimum one: for perceiving such connections where the surnames are different is largely a matter of chance.) Even subtracting the Quakers leaves the total at a still impressive forty-two. Admittedly, in a few cases the connection was rather distant: second cousinhood or cousinhood-by-marriage. In one instance the brother of a member married the sister of another. But even so there were seven pairs of siblings, four pairs of brothers-in-law and three pairs of first cousins. In two cases both a father and his child belonged, in another two both an uncle and his nephew. There was just one instance of both husband and wife.

This high degree of relatedness can be accounted for in part by the tendency for an interest in natural history to run in families. Bean, Bree, Heysham, Masters, Miss Griffiths and Miss Harvey all had had keen naturalists for parents, the three Barnards were descended at one remove from Sir James Edward Smith, Henry Marsham was the great-grandson of Gilbert White's correspondent, Robert Marsham; more remotely still, John Ray even claimed his famous namesake for an ancestor[10] – a lineage to which the President, Gray, laid claim also. In the same way, five members of the Society had sons or daughters who belonged to its successor, the Botanical Exchange Club, while at least five more had well-known botanist descendants. Helping such traditions along, too, was the occasional marriage of a botanist to the daughter or niece of an older botanical friend: Matthew Moggridge and Andrew Bloxam were two who contracted such alliances, anticipating E. F. Linton and H. J. Riddelsdell in a later generation.

Even so there is little sign that kinship influenced the choice of who was elected to the Council and through that how the Society should conduct its affairs. There is no evidence – for example – of the "closely-knit, continually intermarrying, almost dynastic élite" which dominated the Manchester Literary and Philosophical Society for so many years and which resulted in almost five per cent of its membership (and twenty-five per cent of the available offices) being drawn from just six Unitarian families.[11] In so far as people did deliberately introduce their relations, this would seem to have been for no more sinister reason than that these were the most obvious targets for recruitment. Botany at that period was sufficiently fashionable for many families to contain more than one person devoted to it; families, moreover, were in the main far larger than today's and in that more anchored world social life consisted far more of visits to and from one's kin. What more natural therefore than that those familiar with the Society and aware of its attractions should have wished to share it with those nearest to them who happened to share their hobby?

That there was indeed a certain deliberateness behind this striking gathering-in of relations is suggested by the efficiency with which the Society conducted its recruiting in at least one other direction. This was the signing-up of those who had been temporarily thrust into the botanical limelight, most typically as the discoverers of great rarities or of species new to the British Isles. Such people appear too regularly as candidates for admission straight after featuring in the pages of the *Phytologist* for this to be coincidence; and in any case we have the actual testimony of one of them that the Society made an approach in terms she found irresistible. Poor, deluded Mrs Riley, mentioned earlier as a donor of fern specimens to the herbarium, went to her grave convinced that she had been accorded a very great honour in being invited to become a subscriber.[12] Dennes, it seems, had developed the technique of wooing the innocent with flattery which a subsequent Secretary was to deploy to no less effect some eighty years later.[13]

If, therefore, it can be ruled out that the takeover by the social reformers was the result of kin-based power manoeuvring, perhaps it was no more than an inadvertent by-product of Dennes' relentless but over-indiscriminate recruiting. Certainly, the years when the barristers so suggestively moved in were the very ones in which Dennes must have been at his most active. For in 1846 the yearly net gain in subscribers doubled abruptly, continuing at a high level during the two years that followed. Of the eighty-three 'new' ones acquired in that period, however, a sizeable proportion – up to as many as one in six – were either pre-existing members who had allowed their subscriptions to lapse or people who had been taking advantage of the Society's services for a year or two without formally enrolling. This came after a run of years in which resignations were suspiciously few. It thus looks as if in the middle of the 1840s the Society carried out an overdue tightening-up, including a purge of defaulters, and then followed this with a particularly intensive drive to try to make good the losses. It may have been no accident that just about then the Linnean was similarly engaged in a purge. In 1842, having discovered that exactly half of its annual subscribers were in arrears, the Linnean instituted a system of sending out letters of a successively greater sternness, with the threat of legal proceedings as the culminating sanction – and in the event some members did pay up only after having been taken to court.[14] The Victorians and their predecessors had an extraordinary inhibition about harrying members of societies and clubs who failed to meet their debts, apparently because to do so infringed the unwritten code of gentlemanliness. The consequence was that membership lists as often as not bore little relation to financial reality; and if treasurers flinched for too long from correcting the situation, a society could find itself desperately placed, virtually without warning. One near-fatality from this cause at about this very time was the Entomological Society of London. Unfortunately, it managed to keep its crisis quiet, so its botanical sister was deprived of a salutary warning.

In terms purely of numbers gained, however, the Society's recruiting was an uninterrupted success. From the inaugural 38 the total of subscribing members climbed year by year, remarkably steadily, to around 250 by the time reasonably reliable records cease (in 1851).[15] By present-day standards 250 looks hardly impressive, but it was more than the Entomological succeeded in achieving even twenty years later and half the size of the far longer-established and much more prestigious Linnean.

As the membership grew, it also became increasingly diverse socially. Compared with the inaugural period (down to February 1839), when one member in nearly every three had been a medical practitioner, by now other occupations were represented more noticeably. Physicians and surgeons (the latter in the much broader sense then customary, most of them the equivalent of today's GPs) still accounted for one in five, but their dominance was being challenged by clergymen and schoolmasters. Most of the former were Anglican curates or vicars, but the heyday of the clerical botanists was yet to come and there were not as many of these as one might have supposed: in the whole of its existence the Society acquired twenty-nine at the most – as compared with the seventy-four members recruited from medicine. That was not the full representation of even the Church of England clergy, though, for a number of the schoolmaster members were in holy orders too. The schoolmasters, indeed, were by no means a homogeneous group either. At least eight out of the nineteen of these (and one of the women members also) were, or had been, proprietors-cum-principals of small private schools of the kind so characteristic of that period: the 'gentlemen's academies', typically with some twenty to thirty pupils and always financially precarious, which served as the obvious career outlet short of the Church for any man of scholarly inclinations who possessed a modicum of capital. Without capital, the only alternatives for a teacher were the drudgery of an usher – the fate of Hollings, Just and Johns – the uncertainty of private tutoring – tolerable for a foreign visitor like Caspary – or, for the really lucky, the headship of a workhouse school, like Unwin. That is, unless a man was brilliant enough to land a post at a university, like Crouch or Hart or, even more exaltedly, one of the very few professorships of botany or natural history, like Allman, Buckman, Dickie and Mateer.

Apart from this handful of professors the only members who followed botany for their living worked in the offices of museums and botanic gardens or in the secretariats of learned societies and institutes. Despite the pittances that such posts normally carried, there were more of these than might have been expected – and the men were of a far higher stamp than the salaries deserved. One or two had turned professional only after a previous livelihood had failed and had to endure thereby a painful drop in status. But on the whole they were dedicated and content, more than ready to put up with the penury involved in return for the privilege of being paid to devote their working hours to their devouring interest. In effect, they were amateurs who had found shelter from the cold: quite different in their backgrounds and their attitudes from the professional biologists who were later to emerge as the universities expanded from the 1860s onwards. A mere six per cent of the membership at most, they show no evidence of having acted as a coherent interest group; the majority, in any case, could not afford to travel up from the provinces for meetings and came into contact with one another only very little.

Considerably more numerous than the strict professionals were members with occupations in which a knowledge of botany could be rated advantageous. Apart from the medical practitioners of various kinds these included the nine chemists and druggists (just then burgeoning into respectability under the aegis of the infant Pharmaceutical Society), the six nurserymen, the two florists-cum-seedsmen and the nine (or more) gardeners. If the four curators of botanic gardens are added to these last, horticulture as a whole accounted for almost as large a proportion of the membership as the group of botany professionals.

Another substantial group, twelve in all, was made up of booksellers, stationers and printers. One or two of these, Luxford and Pamplin in particular, specialised in botanical literature and so had an involvement in the subject that was close to that of the scientific professionals. Thomas Moore and Francis, who lived largely by botanical and horticultural journalism, fell into a similar category. And the same might be said of Wing and Sly, who relied on patronage from such quarters for many of their commissions as engravers and artists.

All of the rest were engaged in occupations with no apparent botanical connection. These included seven solicitors and eight barristers, a land agent, a relieving officer, a postmaster, eight civil servants, two politicians and seven who worked in banks. There were also five civil engineers, two architects, three accountants, an insurance broker and an estate agent. In short, a good cross-section of what would be reckoned today as the principal middle-class professions.

Industry and commerce were almost as widely represented too. The twelve employed in manufacturing, for instance, made products as various as tiles, nails, wire, paper, tin-plate, time-pieces and furniture. Three of these were heads of substantial firms; another, by contrast, was a humble operative. Brewing and milling each supplied two more; and, complementing these, there were three wine merchants and a corn dealer. Four others were general merchants, while three dealt in timber, one in oil and one in ice. More respectably, there was a commodity broker and a proprietor of shipping. And naturally there were shopkeepers in quite an assortment of sizes: five grocers (together with a grocer's widow), a draper, a draper *and* grocer, a jeweller, a silversmith and, more picturesquely, an umbrella-repairer and an assistant in a tallow chandler's – all of them, noticeably, at the cleaner end of retailing and thus ranking towards the top in its unspoken hierarchy.

Finally, it must not be overlooked that there was another large slice of the membership, not far short of fifteen per cent, that followed no profession or trade at all. Apart from almost all the women members, precluded from paid employment by their social station, there were eight landed gentry, two retired naval officers and a further twenty-two who, like Watson, subsisted on private incomes more or less modestly – though three did so involuntarily, kept by ill-health from pursuing careers. Like Watson too, however, many of them had taken the precaution of acquiring a professional training first (one at the Bar, two in the Church, eleven in medicine), so as a group they were not necessarily all that different in outlook from those who practised a calling.

All in all, it was a social mix that in the terms of the time must have been very much the equivalent to that of the present-day BSBI. But if it was a broadly similar stratum economically and educationally, it differed from its modern counterpart in certain respects very markedly.

For a start, the majority of the members were self-employed. For most of them, time devoted to botany beyond a very sparing point was time taken at the expense of earning: in an all too real sense they gave themselves to the pursuit at no small personal cost. What is more, they did so against a background of unrelenting insecurity. Business was prone to booms and slumps, banks were liable to fail, customers expected interminable credit. There were no pensions or holidays with pay. Financial collapse could mean the workhouse or a debtors' prison.

But the greatest hazard of all was ill-health. A high proportion of all members, rather more than one in ten, died before the age of fifty – and half of these before they were even out of their thirties. Two died within just a month or so of joining, while still in their twenties. Tuberculosis, that great contemporary scourge, particularly of the young, is known to have been the cause of death in at least six cases and must have marred the lives of quite a number more. Epidemic diseases, cholera above all, were great killers too. The members who died young include a high proportion of doctors, who were particularly exposed to infection by their patients. Just as modern medical knowledge would prevent most such deaths now, so it would also greatly reduce the number afflicted with chronic ailments. Woodward, for example, was "almost always in weak health and very frequently in a state of great suffering" from asthma – as was his predecessor as Curator, Henfrey. Thomas Kirk, similarly, was plagued by bronchitis. Lhotsky had recurring abdominal trouble of apparently nervous origin, "with little possibility of being cured". Twining had a leg set badly after slipping on some ice and was ever after confined to crutches.

Even so, if oppressed by ill-health or ruined financially, the Victorians had one great escape route that no longer exists for us: they could emigrate to the Colonies. Eight members chose Australia, the favourite bolt-hole of the era, two New Zealand and one Canada – though most of them went out only after the Society had ended. Several were able botanists and proved valuable acquisitions by the countries they adopted. Kirk, for example, became one of the founding fathers of the science in New Zealand. George Lawson, unable to advance in a botanical career in Britain, rose to similar prominence in Canada. G. W. Francis, another would-be professional, at last realised his ambitions as Director of the Botanic Garden in Adelaide. G. B. Prentice and F. Barnard also made names for themselves in Australian botany.

Even more went out with every intention of returning, their departure overseas being just a step in their careers or merely a protracted sightseeing visit. But if the part of the world they happened to go to was a dangerous one, they all too rarely came back. J. Motley was murdered by natives in Borneo; J. P. Norman was assassinated in Calcutta; T. J. Duthoit died in the aftermath of the Indian Mutiny; T. W. Barlow survived scarcely a month or two in Sierra Leone.

A final difference from today was the much more central place that was accorded to botany by the cultivated. Science was not yet too esoteric and specialised for the average educated person to grapple with, and it still remained the convention to seek to acquire some knowledge of it no less than of art and literature. As a result, even though no longer a single community, the world of science and the world of the arts lay very close together and in some degree intertwined. While some families might be wholly scientific and others wholly artistic, there were still an ample number in which the interests of individual members contrasted or overlapped. It is not altogether surprising, therefore, to find in the Botanical Society several whose shoulders rubbed intimately with those of leading literary and artistic figures. Among these were the father and an aunt of Samuel Butler, the father of Gerard Manley Hopkins, a cousin of Robert Southey (with a mother who had been a friend of Jane Austen), the father of the chief intimate and inspirer of Branwell Brontë, and the headmaster of Edgar Allan Poe (who modelled partly on him the chief character in one of his best-known works, *William Wilson*). Better still, a member who joined in 1850, a Miss Evans of Coventry, was almost certainly the novelist George Eliot.[16] Even in the much tinier Thirsk Club, the Society's immediate successor, the links continued just as strongly, one of its members being the father of George Gissing, another the son of the painter John Linnell and a brother-in-law of Samuel Palmer. It is on the whole unlikely that a botanical society today would prove so artistically well-connected.

All of this makes it easier to understand why the Society's resistance proved so weak to the take-over by the ameliorists. Its consciousness of belonging to a realm set apart was as yet too slight: it did not see Science as an ideal whose purity was in need of being defended. There was no body of professional botanists to speak of, so no move to defend it could come from that quarter either. For members out in the provinces, without the opportunity to meet one another or make the acquaintance of the Officers, it was not sufficiently a social entity; too many of them, in any case, thought only of how much they could extract from it for their own private purposes. From Watson alone was there any clear sense of direction. Consequently, once Watson's arid fare began to lose its savour and no longer proved enough to keep them going, it was only to be expected that the Officers would succumb to the blandishments of incomers who, for all their lack of credentials, had at least the attraction of being positive.

CHAPTER 5

Collapse

At some point during its last few years the Society eventually split. We know of this only from veiled statements made long afterwards, independently, by two who were members at the time:[1] the contemporary literature lets drop not a hint of it and no private papers on or even alluding to the subject appear to have survived. Indeed, the documentation of this final period is as a whole teasingly slight and fragmentary. Such little evidence as exists is mainly negative and open to more than the one interpretation. From now on we are groping through a fog, which for reasons that we can only guess at seems to have been substantially of the Society's own making.

The cause of the split, we are told, was financial. This can surely mean only one thing: that the long-smouldering divergence of opinion between the collecting fraternity and the London 'club men' had finally burst into a conflagration. The funds which the first group saw as increasingly essential to its particular ends the second group was bent on denying. Watson's outburst in the *Phytologist* early in 1849, railing at "the deadlock, towards which some other members are fast forcing it" and calling for the Society to rid itself of "the encumbrance of a herbarium, a library and meetings" in order to make possible the re-engaging of a paid Curator, can hardly be anything but a forceful echo of a debate that was fiercely raging by then on this issue. One obvious compromise would have been a raising of the subscription; but that would have meant abandoning the one feature that the Society had continued to lay emphasis on all through the years, to the point that it had acquired the force almost of a sacred principle. To make the Society considerably more expensive could only have been anathema to that sizeable proportion of the membership which was true to its liberal beliefs before all things.

It was only at the end of the 1840s that the exchange activities had prospered so exceedingly that they had brought matters to this head. Despite the capture of the Council by the ameliorist interest and the consequent change in direction in other respects, this side of the Society had not experienced any counterbalancing swing-down. Quite the reverse in fact: the size of the Distribution in 1848 had been a record, that in 1849 greater still. "Under the able superintendence of Mr Hewett Watson," runs the report for the second of these years, ". . . Mr French and Mr T Moore worked . . . with your Secretary for four whole nights and sent off to the Members about 200 parcels of British Plants." 'Whole nights' was certainly no exaggeration: all three, as Watson pointed out, had "professional avocations to occupy closely their time and attention by day", Moore having recently been promoted to the Curatorship of the Chelsea Physic Garden and French having equally recently acquired a chemist's shop in Pall Mall. They were therefore able to accomplish the task only by "sleep-sacrificing devotedness".[2] In paying tribute to their "immense exertions", the Council expressed the belief that it was "an instance . . . wholly unparalleled for its devotion to Science".

But devotion on so heroic a scale could hardly be counted on twice, so the Society faced the problem of having to raise annually a fresh team of volunteers. The pool from which these could be drawn, though, was narrow: they had to live in or near London, they had to be competent botanists and they had to be firm believers in the value of the Distributions. Ideally, they would have had abundant free time as well; but unfortunately Watson was virtually the only member who could offer this advantage while fulfilling the other criteria – and he had made it plain that he was prepared to do no more now than the comparatively minor task of putting together the packets of non-British specimens for the few members who were interested in receiving these.[3] During the years he had spent putting right the British side of the exchange activities, the foreign material had necessarily had to be neglected; he therefore felt under a lingering obligation to discharge this unfulfilled part of his responsibility.

By 1851 Watson had apparently dropped even this; for along with the names of the now customary two volunteers to work alongside Dennes there appears in that year a third: that of G. C. Churchill, a Manchester solicitor and keen mountaineer who specialised in the flora of the Alps. Presumably he had turned out to be willing to devote part of one or more London visits to advising on the foreign material to be selected – with first choice for his own herbarium, no doubt, as a sizeable inducement.

That the Society had to go so far afield to find a substitute for Watson's expertise in this quarter was not a little ominous. It was further ominous that one of the other two volunteers that year was J. A. Brewer, who lived far out in mid-Surrey and was insufficiently well-off to own a London *pied-à-terre*. Clearly, the larder of appropriate candidates was already almost bare and the position was becoming desperate. As even the least sympathetic Council member could hardly fail to realise, now that every second person in the

Society was requiring to be sent an annual parcel the work involved was simply too formidable for voluntary effort alone to be relied on any longer. Either the number who shared in the Distributions would have to be cut back – and such a course, again, would have been repugnant to the liberals – or the expense would have to be incurred of a part-time paid Curator.

It is at this point, early in 1851, that the show-down which Watson had been predicting seems most likely to have occurred. The previous summer and autumn had witnessed a steady stream of papers on ameliorist topics; after that date, and throughout the remaining years that followed, British field botany appears to have formed the fare at the meetings to the exclusion of anything else. Up to 1851 a significant proportion of the new members elected yearly were people who are not known to have been collectors; after 1851 only a single such person can be identified. From around this time, too, all but one of the new people elected on to the Council who can be said to have had some standing as field botanists – six of them altogether – have one other feature intriguingly in common: they were all men who had moved to live in London very recently, even within a matter of months. The only explanation for this phenomenon that suggests itself is that the field botany faction was short to the point of embarrassment of supporters with the requisite qualifications who were available for putting forward. It may well be that it was engaged in seeking to dominate the Council now in turn, just as its antagonists had done so successfully some years before. Or it may be that the opposing faction had suddenly caved in, some resigning and others merely choosing to absent themselves (for three of them continued to be at any rate listed as Council Members until the Society's end), leaving Watson and his allies to scrape together substitutes from their side as hurriedly as they could. Certainly, it would not be at all surprising if just around this time the advocates of botany-for-social-betterment had eventually lost heart for further struggle. For by 1850 the tide of reform everywhere was on the turn. After only very partly succeeding in their aims, the ameliorists had found their political progress blocked and were abandoning their very varied fleet of vehicles. Soon even their arch-exemplar, Chadwick, would be falling from governmental grace.

Whatever the inside story, it was the field botanists, quite unambiguously, who won. For 'to the victors the spoils' – and the spoils took the very concrete form of, at last, a paid Curator once again. Interviewing to this end went ahead in June of 1851; and "from among several candidates" (according to the minute-book) a young civil engineer from Scotland was selected. This was John Thomas Syme. He was twenty-seven by then, bored with building railways and with the prospect of inheriting in due course the ancient estate of

Balmuto in Fife – though he was very far from being well-off. Syme was a singularly lucky catch for the London Society, for with a well-known natural history artist for a father and two maternal aunts who were keen Scottish plant-hunters, he had been close to the botanical purple more or less from birth. The Edinburgh society had enrolled him when he was but fifteen, and in the years since then his field reputation had ripened and led to his election as its honorary Curator. It had been his very first paper read before that society, on the flora of Orkney, in February 1850, that had caught Watson's talent-spotting eye and given rise to the suggestion that he try putting in for the London job. Glowing testimonials from the leading Edinburgh members must have made his candidacy virtually a walk-over.

By now, thanks to its very much larger membership, the Society could afford to pay double what it had felt able to offer Henfrey and Edmondston.[4] In return for this, Syme was required to be on hand every Monday, Wednesday and Friday, from ten till five. For the rest of the week his time was his own; and those first few summers he chose to spend a great part of this acquainting himself with southern England, amassing specimens for his own as well as the Society's herbaria with what seems today a horrifying abandon. Within a year or two he was lecturing regularly on botany at two of the London hospitals and was firmly ensconced as one of the small handful in the Capital who were able to make a living from the subject.

At the end of his first full year in office the Society elected him a member of the Council, in an ordinary capacity. This was decidedly strange; for it meant that, in effect, the employee also became one of his employers. Henfrey, however, had been treated similarly, so it must be assumed this was the Society's recognised way of showing its respect for its Curators. Although the occupants of the post could equally well have been admitted to the inner counsels on a special, ex-officio basis, the preference was evidently for making no distinction between them and anyone else of Council calibre. At the time when Henfrey had been elected, however, the precise composition of the Council was a matter of little moment. Now, though, it clearly made quite a difference that the one person in the Society most intimately concerned with the Watsonian side of its activities had a voting hand in the framing of its policy. This is thus further evidence that the field botanists had become by then the dominant power on the Council.

One year later this power was demonstrated forcibly. An announcement was made that the Council had decided to restrict the Society's general operations to the British flora alone, in recognition of the fact that it was only in this that the majority of members had an interest. To meet the needs of those who wished to receive non-British material, a new autonomous section had been formed, under the name of the Foreign Exchange Club. Any member of the Society could join this on completing a special form and sending it in accompanied by sixty postage stamps. The Club would be funded by means of a charge on the specimens sent, ranging from a penny to threepence depending on their degree of desirability (an interesting throwback to Watson's very earliest exchange ideas). If the money thus raised proved to be inadequate, the members would be called on to make up the deficiency themselves, up to an amount per head not exceeding five shillings in any one year. If that still proved too little, the Club would be dissolved. The Society's Curator was to conduct all the Club's operations, but only "at such times and in such a manner as may least interfere with the business of the parent Society".[5]

The brusque, no-nonsense tone is symptomatic. Watson and his fellow spirits were not being allowed to chop off the branches that they regarded as superfluous, but as the price of retaining these they were able to insist on their own exacting terms. It was not very much further on from that to the cleanly trimmed-down trunk that Watson had been calling for. As things were going, in only a year or two more, that desired end might indeed have been reached, without the constitutional convulsion that in the event was to effect it.

Whether the Foreign Exchange Club ever did start up we do not know. But if it did, its existence can have been only extremely brief. For within two years the Society itself was being pronounced extinct.

The end seems to have come with a dazing suddenness. There are no hints in the letters of Watson and Syme of this period that the Society was in difficulties or that any reining-back, much less closing-down, was being planned or even so much as talked about. The Distributions had been continuing year by year, with seemingly as many participants as ever; and though it was Watson and Syme who latterly provided the bulk of the specimens, Baker (who joined in 1851) is known to have contributed largely too and he may well not have been the only one. A report on the Distribution of 1855 by Syme mentions that some "highly interesting" plants had been received that year, one of them the long-lost Holy Grass, *Hierochloe borealis*, sent from Caithness by Robert Dick, the Thurso baker and hero of Samuel Smiles.[6] One gathering at least of Syme's exists made in October 1855 and bearing one of the Society's labels, which implies that there was also a Distribution in 1856. True, there are no reports of any meetings having taken place after February 1853, but reporting of these had been patchy for two years by then and this does not necessarily therefore point to their discontinuance.

All the same there are some puzzling features which seem to suggest that things by then were not entirely normal. Syme, for example, in one of the very few letters of his that have come down to us, reveals that in the spring of 1855 his visits to the Society's rooms were sufficiently infrequent for mail sometimes to take as long as a fortnight to reach him.[7] Why was this? He had certainly not resigned, for he is still named as the Curator in a circular late the next year. But it could well be that he had asked to be relieved of his duties temporarily, in order to devote all his time to preparing the course of lectures that we know he was due to start giving in the spring of the next year.[8]

There had occurred, too, a marked deterioration in the keeping of the minutes. As early as 1848 the entries in the minute-book become noticeably more perfunctory, more illegible and more prone to slips, suggesting that Dennes was operating under some kind of stress. For a time a different hand even takes over. After the Anniversary Meeting of that year there is a five-month gap in the record, followed by a further, six-month one soon after. For the whole of 1850 and 1851 together only five meetings are entered up, though certainly more than this are known to have taken place. After November 1851 the minute-book peters out altogether.

Whenever a Secretary is under pressure, the writing-up of minutes is usually the first of his duties to suffer; so ordinarily it would be wrong to read all that much into a lapse in this particular quarter. However, at around the same time the reported membership figures also start looking suspect (due, apparently, to under-reporting of resignations), which suggests that the decline in efficiency was affecting the Society's record-keeping more generally.

It may be no coincidence that 1849 was also the first year in which Dennes was left on his own, free of Watson's close supervision. If, later, Syme was absent for much of the time too, then the Society was abnormally dependent in these final years on the conscientiousness of its hitherto almost excessively hard-working Secretary. Indeed it was doubtless the very devotedness that Dennes had shown up till then, combined with his long experience, that left everyone with no qualms about leaving the running of the Society effectively to him alone.

On the surface he continued to be as diligent as ever, at least up till the time of the last reported meeting (in early 1853). The reports still appeared over his initials and he regularly features among the donors of specimens. After that point, however, the silence that descends on the literature may genuinely be a signal of the withdrawing of his attentions, confirming the claim of Irvine that the Society "died of sheer atrophy or inanition".[9]

All that is sure is that Dennes was the one who received the blame for the chaos that eventually came to light. This was in the summer of 1856. By the beginning of that October the financial position, it had been discovered, was so irretrievably grave that word went out that there was no alternative but for the Society to be dissolved.[10] Watson confided to Babington: "Dennes has managed that Soc^y. pretty much as he has managed his own affairs; and with a similar result. He usurped the functions of Treasurer, but failed to perform the duties. No accounts or receipts can be got from him; and the Society is found to be far behind with rent, etc."[11] Years later, he was to assure Balfour too that it was "the financial mismanagement and personal vanity of Dennes" that led to the Society being "irretrievably ruined".[12]

There is no reason to suppose that Dennes had acted dishonestly: if he had, he would surely not have received the parting donation from the members that he did. Without question, though, he had acted dishonourably; for clearly his correct course as soon as he found himself no longer able to cope would have been to resign his office. That he had failed to do so must be put down to the intensity with which he had come to identify with the Society. Without it, his life would have been emptied of much of its purpose. By a tragic irony he was a victim of his very devotion.

On the twenty-fourth of November a meeting was called of the Resident Members. We only know of this thanks to the lucky fact that Watson used as scrap paper the reverse side of copies of the circular which reported what transpired.[13] In essence, this was that the following resolutions had been passed unanimously:

1. That it is expedient to dissolve the Botanical Society of London, as at present existing; reserving to its Members the right of re-constituting themselves into a Society bearing the same name, and for the same general objects.
2. That the Books and other property of the Society shall be sold, in order to raise a fund for paying the arrears of rent and other outstanding claims on the Society.
3. That Mr. Reynolds *(Treasurer)*, Mr. Syme *(Curator)*, and Mr Hewett Watson are appointed a Committee to carry out the two preceding Resolutions, and to wind up the affairs of the Society, in such manner as they shall find expedient.
4. That the foregoing Resolutions be printed, and a copy be forwarded by post to each of the Members of the Society whose addresses can be ascertained.

There followed a series of questions designed to remedy the lapse in the membership records, and the circular then concluded by asking bluntly: "In the event of the Botanical Society being re-constituted for the promotion of

20 Bedford Street, Covent Garden:
the home of the Botanical Society of London for most of its existence.

Botanical Society of London.

Instituted November 29th, 1836.

75, NEWMAN STREET, OXFORD STREET.

OFFICERS FOR 1838.

President.
JOHN EDWARD GRAY, Esq., F.R.S.

Vice-Presidents.
Dr. D. C. MACREIGHT, F.L.S. — CHARLES JOHNSON, Esq.

Treasurer.
JOHN REYNOLDS, Esq.

Curator.
DANIEL COOPER, Esq., A.L.S.

Secretary.
GEORGE E. DENNES, Esq.

Other Members of the Council.

Francis Bossey, M.D.
Edward Charlesworth, Esq., F.G.S.
W. M. Chatterley, Esq.
Joseph Freeman, Esq.
Thomas Webb Greene, Esq., B.C.L. F.H.S.

Æneas MacIntyre, L.L.D. F.L.S.
W. H. Ranking, M.D.
James Rich, Esq.
W. H. White, Esq.

Local Secretaries.

Mr. Wm. Baxter, A.L.S. Botanic Garden, Oxford.
Thomas Bodenham, Esq., Hanley, near Shrewsbury, Salop.
J. A. Brewer, Esq., Reigate, Surrey.
James Buckman, Esq., Cheltenham.
Charles Conway, Esq., Pontnewydd, Monmouthshire.
Dr. Ferdinand Krauss, Cape of Good Hope.

H. B. Fielding, Esq., Stodday Lodge, near Lancaster.
F. C. Lukis, Esq., Guernsey.
Edwin Lees, Esq., F.L.S. Curator, Nat. Hist. Soc., Worcester.
Roberts Leyland, Esq., Halifax, Yorkshire.
Dr. Bell Salter, F.L.S. Poole, Dorsetshire.
Arthur Wallis, Esq., Chelmsford, Essex.

The Botanical Society of London is instituted for the promotion and diffusion of Botanical Science, by the formation of an Herbarium, the Exchange of Specimens with other Societies, or with Individuals, the reading of Original and other Papers, the formation, also, of a Library and Museum, and by the establishment of a Botanic Garden, as soon as the Funds of the Society will permit.

Title page of Annual Report for 1838.

A CATALOGUE

OF

THE HERBARIA

AND

LIBRARY

OF THE

BOTANICAL SOCIETY OF LONDON,

Removed from the Rooms,

BEDFORD STREET, COVENT GARDEN,

In consequence of the Dissolution of the Society.

Which will be Sold by Auction, by

Mr. J. C. STEVENS,

AT HIS GREAT ROOM,

38, *KING STREET, COVENT GARDEN,*

On MONDAY, the 9th day of FEBRUARY, 1857,

At One o'Clock precisely.

May be viewed on Saturday and Morning of Sale, and Catalogues had of Mr. J. C. STEVENS, 38, King Street, Covent Garden.

ALFRED ROBINS, Printer, 7, Southampton Street, Strand.—W.C.

A CATALOGUE.

On MONDAY, the 9th day of FEBRUARY, 1857,

At One o'Clock precisely.

PLANTS, &c.

LOT		PARCELS
1	Duplicates of British plants, named	9
2	Herbarium of British flowering plants, fastened to paper, about 1200 species, all named	12
3	British ferns, on paper, very complete set, and all named	1
4	Twenty very rare British plants, including Arenaria Norwegica, Orchis hircina and Spiranthes cernua	1
5	British Rubi, named by Bloxam, Leos, &c., 8; Bloxam's fasciculus of Rubi and Leighton's Shropshire ditto, 2	
6	British Hieracia, on paper, and Baker's Hieracia of North Yorkshire, named	10
7	British Roses, on paper, 1, and British Saxifrages, ditto, 1	2
8	Leefe's Salicetum Britannicum	2
9	Two sets of British plants, collected near Thame and Embledon, &c., about 600 species, named	1
10	Two sets of plants, collected near Malvern and in Lancashire, about 800 species, named	5
11	Berkley's British fungi, fasciculi, 1, and fungi, from Bristol, 2, named by Stephens and others	2
12	British mosses, about 150 species, duplicates	3
12*	British lichens	3
13	Ditto Algæ and Hepaticæ, about 60 species	2
14	Medicinal plants, on paper, about 100	1
15	European plants, named, about 1000 species, many duplicates	11
16	Ditto, including rare species from Transylvania, about 300 species, many duplicates	6
17	Wirgen's Herbarium Monilarum Rhenanum	1
18	Plants from Mexico, &c., collected by Bates and Siebold, 10*	5
19	Miscellaneous exotic plants, named, about 300	2
20	Exotic ferns, from Riley, named, about 100 species	1
21	Ditto mosses, lichens, &c., about 50 species	3

Page from the catalogue of the auction of the Society's effects in 1857.

British Botany, are you willing to continue your membership in the renovated Society?" A further circular was promised on how the resolutions had been followed up; but if this was ever issued, no copies of it seem to have survived.

The careful wording of the resolutions betrays the legal purpose that lay behind this meeting: the formalities of an official winding-up necessarily had to be gone through if any successor body was not to be encumbered with the Society's debts. That Watson and Syme should have gone to such lengths is testimony enough that the talk of starting a new society was very much in earnest. And that this was to be at last the slimmed-down, streamlined model that Watson had been advocating for so long is confirmed in their correspondence. It was to be "merely for distribution and making an annual report on British Botany", Syme wrote that January to Sir William Hooker.[14] Despite Watson's show of agonised breast-beating there seemed a good prospect that the disaster was now going to bring him exactly what he had been wanting all along.

The Society having been dissolved, there remained only the melancholy business of disposing of its effects. The library and herbarium were duly auctioned by Stevens on the ninth of February, in 253 lots. A ten-page catalogue (a copy of which still exists)[15] ensured a good turn-out from the botanical world, including some at least of the keener ex-members. Watson himself made a point of trying for some of the collection of *London Catalogues* marked up for particular areas, but in at least one case subsequently discovered that it was the author against whom he had been vainly bidding.[16] A Manchester collector, John Hardy, who had latterly taken part in the Society's Distributions, is said to have bought part of the herbarium;[17] but certainly the greater part of this, along with the remaining stock of British duplicates and "other lots", was purchased by F. Y. Brocas, a young Hampshire botanist who saw this as the opportunity to set up in business as a dealer in herbaria. With commendable public-spiritedness Brocas began by keeping the collection together and making it available for general consultation (doubtless for a fee) as 'the London Herbarium of reference';[18] but in due course he tired of this and must be presumed to have dispersed it piecemeal, for not even portions of it can now be traced. Some at least of Brocas' own personal herbarium, however, has recently come to light in the cellars of the Horniman Museum, in South London, and incorporated with this is what must surely be the collection formed by Ayres around Thame, in Oxfordshire, and donated to the Society in 1843–4. Almost certainly this was one of the 'other lots' which Brocas purchased at the auction, and it may thus constitute the sole tangible relic of the sale that has identifiably come down to us.

The auction produced a surplus of about £30 after the clearing of all the debts. With the consent of those members who had attended the winding-up the whole of this was given to Dennes, who by that time, according to Watson, "was really in a starving condition".[19] From these words it would appear that some private financial catastrophe, probably an unwise speculation which had gone badly wrong, was the root cause of the Secretarial distraction which had brought the Society down. To try to retrieve the position, Dennes was driven to the length even of abandoning his profession, taking his name off the roll of solicitors, voluntarily, sometime during 1858. The next year the London street directory shows him living at an address in Queen Square. By 1861 he had been taken off the books of the Linnean and Edinburgh Societies, and legend has it that he had sought refuge in Australia.[20] It must be feared that the rest of his life was one of penury as well as oblivion. It was a doubly sad ending for the one person who toiled hardest to make the Society a success, who had nursed for it the highest ambitions and who had finally had to witness all those many years of effort seemingly come to nothing.

All hint of the true reason for the Society's demise was suppressed. Contemporary periodicals report only the auction and the bare fact that the Society had ended – though it is hard to believe that the inside story was not known to some at least of their editors. Even the Society's own members did not necessarily learn, or even come to suspect, that there had been anything untoward. Over half a century later, by which time there can surely have been no need to hide the truth any longer, J. G. Baker had no better explanation to offer for the Society's collapse than that funds had been found to be insufficient to pay for a Curator and to keep up rooms in central London – and the Society had therefore to be dissolved, he concluded lamely.[21] Baker, though, had been only a recent recruit at the time and an unworldly youth, living far off in the North, is unlikely to have been privy to very much. But the same can hardly be said of Gray, merely nominal President though he seems to have become. His explanation, while quite different, is no less of a distortion of the facts: "After several years, [when] the Society seemed to have done its work of distributing well named specimens, opportunity was taken of the death of several of the more active members and the removal of others from London to dissolve it."[22] A careful scrutiny of the membership records, defective though they are, reveals no evidence to support either of these two assertions. And the fact that the Distributions continued for a further century hardly lends credence to the first one either. Gray must surely have known, at least in broad outline, what had really occurred, but either he had determined to stay reticent or, having learned both versions of the events, it was the official one alone that lay lodged in his memory in those last years of his life.

Even the very fact that the Society had ceased to exist was surprisingly slow in becoming known. The scurrilous Irvine, who would certainly have been among the first to crow, was clearly unaware of the winding-up when he penned an editorial preface to his new series of the *Phytologist* that January: the reborn journal, he assured his readers, had "not the most remote intention" of superseding the Society "as a medium of public usefulness". If a Londoner and a committed enemy of the Society like Irvine had not heard the news by then, it must have been indeed closely guarded.

Much less surprisingly, the news took a very long time to circulate overseas. For at least three years afterwards letters from foreign botanists were still arriving addressed to the Society – and were being delivered, with an irritating frequency, to the rooms of its near-namesake in Regent's Park.[23] It was a just revenge, from beyond the grave, for the heedlessness with which the late-coming Royal Botanic Society had chosen to inflict upon the world this ambiguity of titles.

But even though in the eyes of the postal authorities the two societies were now one and the same, in the eyes of the scientific world there was no question whatsoever of the survivor filling the place of the deceased. They had developed into wholly different institutions and it was impossible for what was by now a purely horticultural body to cater for the interests of British field botanists. Unfortunately, though, there was no other society in the metropolis concerned with botany exclusively which could function as an orphanage – nor anywhere else in the land either, closer at any rate than Edinburgh, which could have offered itself as an alternative base, as Liverpool was to do, in parallel circumstances, for the astronomers. For generations to come London's botanists consequently had little choice but to join the much more broadly biological Linnean – always supposing that that society would admit them – and try to make the best of its still-continuing limitations. The effect of this was hardly healthy for the Linnean, for it tilted even more disproportionately in a botanical direction a body already seriously imbalanced by the hiving-off of the Zoological and Entomological Societies. C. Davies Sherborn, the bibliographer, was fond of recalling one occasion at the Linnean, many years later, when the opening remarks of a zoologist speaker were completely drowned by the noise made by the botanists departing from the meeting-room as soon as the lights were turned off for the first of the lantern-slides. "It ought to be taken over by the botanists, who need a society of their own," Sherborn is said to have observed more than once.²⁴

The loss to British botany, however, went deeper than that. For all its defects the Botanical Society of London had served to bring together in a single forum specialists in many of the different aspects of botany: in physiology as well as taxonomy, in all the cryptogamic orders as well as the higher plants, in the vegetation of exotic areas as well as that of Britain. In time it might thus have evolved into the all-embracing counterpart of the Entomological Society of London, providing a unifying focus for the science when the onset of full-scale professionalism later began to pull botany apart into a series of specialised sub-disciplines. For such a truly national botanical society in the fullest sense the Linnean, with its inherently divided loyalties, could never hope to serve as an entirely adequate substitute.

The collapse removed from the scene, moreover, the one substantial association of botanists which had begun to give promise of acting as a national network capable of undertaking co-operative research.²⁵ In effect, this was the shape, albeit an ill-formed and rudimentary one, into which Watson had successfully been hammering it. The volumes of the *Cybele Britannica*, that grand summation of knowledge on British plant distribution as it stood at mid-century, embodied the collective achievement of the Society as well as the fruits of Watson's own observation and thinking. And his introduction in the third and final volume of that work, in 1852, of a new and less cumbrously large mapping unit, the 'vice-county', had now provided the army of local workers with subdivisions of the countryside with which at last they could readily identify individually. Henceforward the making and logging of additional records for the 153 of these was to form the main driving-force for British field botany, for the greater part of a century.²⁶ Watson's accountant-like approach to the study of why plants grow where they do was not, as he fondly supposed, the high road leading to what we recognise today as the science of ecology. However, he had done the subject a great if unintended service by providing the typical amateur with a clearly perceptible goal, which could be seen as modestly useful scientifically as well as deeply satisfying aesthetically. For as the geologists had discovered already, there is no activity with a more immediate appeal to the field man than the pioneer working-out of distribution patterns.

Watson's success in enlisting and drilling this large, country-wide corps of volunteers had already given rise to at least one instance of would-be imitation. At a meeting of the Society in June 1850 Arthur Henfrey made the interesting proposal from the chair that six to twelve members, say, residents in parts of the country as distant from one another as possible, "should form an association for cultivating critical plants". The efforts of these, he suggested, should be co-ordinated by a botanist resident in the metropolis "who should receive from any source specimens of plants to be carefully preserved, together with seeds to be distributed to all the cultivators". In this way named entities of doubtful validity could be tested by exposure to a range of climatic conditions. The preservation of voucher specimens, he added, "would render all the observations at once applicable as exact scientific evidence".[27] But despite his offer to act as the London co-ordinator himself, there appears to have been no response to this scheme for comparative experiments in genecology at such a remarkably early date. Probably unbeknown to him, however, only a year or two later one of the formerly more active members, David Moore, did go so far as to test the distinctiveness of one of the Irish Sea Lavenders of the *Limonium binervosum* group by growing it in the Dublin Society's garden at Glasnevin.[28] Another, Buckman, was to embark on controlled transplant experiments at the end of the 1850s while Professor at the Royal Agricultural College, Cirencester, receiving backing from the British Association for his work. It would be nice to think that both had been inspired by Henfrey's abortive brainwave; but, alas, there is no evidence of this, either one way or the other.

Apart from the main mass of records on which Watson's successively weightier compendia were able to be built, the Society left one other major legacy which has tended to be overlooked. This is a stratum of unexpectedly rich and accurately named material in the herbaria of many of today's regional and local museums. At the time when the Society was founded a chain of scientific-cum-literary institutes, such as the Plymouth Institution, the Portsmouth Athenaeum and the Royal Institution of South Wales, was springing into being up and down the country to serve as the places for the local learned coteries to meet and to house the collections of books, memorabilia and reference specimens which these were very rapidly amassing. A remarkable number of the Botanical Society's members prove to have been the natural history or botanical curators of these bodies, in either a paid or voluntary capacity, and the prime motive for their belonging was clearly to take advantage of the Distributions to render the corporate herbaria in their charge more taxonomically sophisticated and more nationally representative. But the greater these collections grew, the more intolerable the burden they became to their private shareholders or the local associations in which their ownership was vested. One by one, therefore, from as early as 1846, following the passing of the Museums Act in the previous year, they were handed over to the ratepayers, no doubt to the considerable relief of the former owners, and so went to form the nuclei of the subsequent municipal collections. To a far greater extent than is generally realised those invaluable arrays of early voucher specimens, the solid bedrock on which many a present-day local Flora rests, thus owe their existence and their continuing usefulness to the great dissemination mechanism devised and perfected so many years ago. The Botanical Society of London may long since have vanished into the dust of history, but it has left behind this splendid set of relics to bear witness to the industry and dedication with which it carried out its central task.

PART TWO

The Botanical Exchange Club

Victoria regia

Floreat Flora

CHAPTER 6

Interlude at Thirsk

"The London Botanical Society . . . is not defunct, but in abeyance, shortly to be reconstituted, and to be rendered more efficient than ever." So wrote the reviewer of the fifth edition of the *London Catalogue* in the *Phytologist* in October of 1857,[1] confidently repeating the assurance that Watson and Syme had taken the opportunity to give in their preface to that work. That preface must have been written in the early months of the year, however, for by the time the review came out the idea of reconstituting the Society in London had quietly been abandoned.[2] It is not clear why. Possibly Watson and Syme had pinned their hopes on one individual in particular, who had promptly let them down. Or they may have envisaged shouldering the task themselves, only to find after all that their commitments would not allow this.[3]

Luckily, Baker in the meantime had come to the rescue. As one of the keenest of the former exchange distributors, he was loath to see this invaluable service interrupted and accordingly volunteered to keep it going until the expected new London body was off the ground.[4] Having recently helped to found a small natural history society in his home town of Thirsk, he was able to put this forward as a convenient ready-made harbour where the national ship could safely lie up while waiting for fairer weather.

John George Baker was at that time the coming young man of British field botany. Three years earlier, recognising his promise, Edward Newman had tried to interest him in taking over the editorship of the *Phytologist*, following the death of the long-serving George Luxford;[5] and shortly after that he must have won the heart of Watson by publishing a thirty-page essay on the geological factors influencing plant distribution.[6] A product of that great nursery of Quaker naturalists, Bootham School, where at even as early as thirteen he had been put in charge of the herbarium, he was now helping his father to run a large general drapery and grocery business with premises in the Market Square. Although a small country town in the North Riding of Yorkshire was hardly the obvious base for an organisation with national pretensions, there was certainly no denying that Baker was eminently well-qualified for the work that he was offering to undertake.

J. G. Baker.

That November of 1857, accordingly, at its fifth Annual Meeting, with Baker in the Presidential chair, the Thirsk NHS passed a resolution establishing a society-within-a-society "for the interchange of dried specimens of British Plants, especially of the higher orders" – these last words because, although it had been intended to restrict it just to flowering plants and ferns, a keen bryologist member, J. H. Davies, had persuaded them to extend the scope to mosses.[7] Respective Curators for these were appointed, in the persons of Baker and Davies, and a new category of Corresponding Member created to empower non-locals to join without having to pay the entrance subscription. Thus, in admirably flexible style, as the 'Thirsk Botanical Exchange Club' (to use the inverted commas of many contemporary references to it) the ghost of the London Society entered upon what was to prove – contrary to Baker's innocent supposition – a visit of as long as eight years. For the Club speedily attracted to itself the cream of the country's exchangers, most of them, so far as one can tell, ex-members of the London Society, and Baker ran it so efficiently that everyone soon lost interest in pursuing any other initiative. As if to proclaim the legitimacy of the succession, Watson presently even handed over to it the London Society's records, such as these were, and the remaining stock of that body's solitary volume of *Proceedings* "for distribution among the members".[8] Thus there was an essential continuity between the two, sufficient to entitle the BSBI to regard its ancestry as descending from the BSL, albeit imperfectly.[9]

Like its predecessor, the Club relied at first on the *Phytologist* to carry regular reports of its meetings and notes on the specimens distributed. But like the Club itself, this had in the meantime experienced a reincarnation: in 1855 William Pamplin, the leading botanical bookseller and herbarium agent, had agreed to take it off Newman's by then sagging shoulders and had installed as editor the excessively opinionated Irvine. Under Irvine it became much less of a scientific periodical, increasingly unreliable in what it published and to a flagrant degree a means by which to snipe at Watson. After five years of this Watson – who was later to dismiss it, with contempt as "a journal beneath question"[10] – could tolerate the situation no longer and started planning a new monthly to compete with it and, with luck, put it out of business. This was to be called the *British Botanist* and was to be edited by Syme.[11] He would have preferred "a man of more active and suggestive mind", Watson admitted, but Syme had the essential asset of being resident in London; besides he was "a man very unlikely to give offence, having nothing of any *waspishness* about him".[12] Watson fully expected the costs to be double the returns and to make up the difference out of his own pocket.[13] Extensive canvassing produced promises of support from most of the country's leading botanists, including Hooker, Babington and Baker. The first of two printed circulars (a copy of which survives at Kew) announced it as "intended chiefly for records in British Botany, descriptive, critical and topographical" and as "taking rather higher scientific ground than the original *Phytologist*" – this last a pointed dig at Irvine. Had it materialised, it would thus have been exactly the journal tailored exclusively to its interests which British field botany was so conspicuously to need and lack for almost another century, a lack which was only really met in 1949 with the launching of *Watsonia*. Most unfortunately, though, what Watson was to describe, mysteriously, as "untoward circumstances" intervened and caused the plans to be put off indefinitely.[14]

Additionally fatal to those plans was the founding in the meantime, in 1863, of Berthold Seemann's *Journal of Botany*. Conceiving it as an English-language counterpart to his German *Bonplandia*, Seemann originally saw this as giving over a full half of its contents to British botany and set out to steal many of Watson's prospective subscribers.[15] When Baker subsequently accepted Seemann's invitation to assist with its editing, Watson was made to feel even further ill-used, for he always looked upon Baker as the foremost of his protégés and indeed retained a special regard for him to the end, paying him the ultimate honour of choosing him as his executor.

Thus, very sadly, a marvellous chance was lost. By the bringing together of the activist remnant of the London Society now clustered around Baker and the readers of a journal exclusively devoted to the advancement of knowledge of the British flora, a much more thoroughgoing national body might have crystallised afresh. As it was, the *Journal of Botany* was left with a monopoly and, though the Thirsk Club and its successors were to use it as the medium for their reports, all through the years British botany was seldom to be very prominent in its pages and it failed to win much loyalty from the great majority of amateurs. As a result, after almost a decade in existence, it was still managing to sell only some 250 copies an issue[16] and was passing from one publisher to another with a frequency that appeared more and more ominous.

A further hint of what-might-have-been was the founding in London in 1862 of a Society of Amateur Botanists.[17] The demise of the BSL had left an acheing hole in the botanical life of the Capital which was felt especially acutely by novices. A few years earlier M. C. Cooke, a leading mycologist and at that time headmaster of a school in Lambeth, had started taking some of his pupils on country rambles; gradually these had attracted outsiders and Cooke was encouraged to build a formal field club around them. The prospectus of this had a familiar ring: the aims were "to be of mutual assistance in the study of British plants by organized excursions, by the establishment of a herbarium, museum, and library, the interchange of specimens, the communication of papers, and such other means as from time to time might present themselves." To any reader of those words who had known the Botanical Society of London it must have seemed that that body was about to undergo a second birth – and once again, by a neat coincidence, after a conception in Lambeth. What emerged, however, proved much more modest. The number who joined over the next two years amounted to only about fifty, so it was possible for the meetings to take place in a room which Robert Hardwicke, the natural history publisher and bookseller, was glad to make available at his shop in Piccadilly. These, on alternate Wednesday evenings, were supplemented in summer with excursions on Saturday afternoons. With no premises or staff to have to pay for – on the new field club model just then becoming popular – the subscription needed to be but slight, and at half-a-crown the cost of joining was a mere eighth of that of the Society's vauntedly inexpensive predecessor. Had the BSL only been founded twenty-five years later, it too might well have sidestepped the burden of property altogether like this. On the other hand the new body failed to realise its intention of organising exchanges and its herbarium, likewise, never came to much. Where it did succeed conspicuously was on the field front: for its early plan to compile a register of all plants found in the London area by its members had matured by 1866 into the preparing of a full-scale *Flora of Middlesex*, the publication of which three years later by the Society's patron, Hardwicke, set a new high standard for county Floras generally. The nominal authors of this, Henry Trimen and W. Thiselton Dyer (a future Director of Kew), had been the Society's leading spirits; and they were massively assisted in the work, with his usual self-effacing anonymity, by a fellow member over twenty years their senior, Revd W. W. Newbould.

Before this Society petered out, after only a brief existence – though, in a sense, it continues to this day, in the guise of the Quekett Microscopical Club[18] – Trimen, Dyer and Newbould had all enrolled in the Thirsk Club as well. So too had one at least of the sprinkling of non-London members the Society had managed to attract: the Manchester-based Charles Bailey. It was thus from this source that the Thirsk Club's successor, the Botanical Exchange Club, came to acquire two who were to be among its staunchest and longest-serving Secretaries. For that, if for nothing else, the Society of Amateur Botanists, ephemeral though it unhappily proved to be, deserves to be gratefully remembered in these pages.

How long Thirsk would have remained the centre of gravity of field botany in Britain can only be a matter for conjecture. At some point Baker must surely have tired and then no doubt the Club would have found some alternative base.

But of that there was still not a hint after six and a half years when, with even greater suddenness, disaster struck again.

What happened is best recounted in the graphic words of one of its later Treasurers:[19]

> On the night of the 9th of May 1864 I was sleeping, for some reason, at the house of my grandfather in an adjoining part of the town, when we were roused up at two o'clock in the morning by the thrilling tidings that the Baker premises were on fire. My uncle ... immediately dashed to the scene, and, to my everlasting gratitude then and ever after, took me with him, I being then six years old, and deposited me at my parents' house where from the windows I could look across the corner of the square at the blazing pile. No wonder the scene is burned into my remembrance, as vivid now as it was then, near 70 years ago. The whole large block of building a mass of flames; the crowd in the market place; the string of buckets passed by hand from the mill-dam over 200 yards away. This most energetic performance was about as much use as a child's squirt would have been. The town fire-engine, a real museum-piece, had given up the ghost early in the day – or night – and the fire engines summoned from York and Darlington, each about 24 miles away, naturally arrived far too late. Imagine the effect on a little boy's mind. One moment I could see through the reddened windows the familiar chairs and furniture of the up-stairs drawing-room I knew so well; the next they had disappeared into the inferno as the floor gave way. The fire spread upwards from the basement, started, it was supposed, by a match dropped down the cellar grating by some smoker returning late from a dance held in the town that night.
>
> Gilbert Baker and his family had the narrowest of escapes. Really they owed their lives to the infant Edmund, who had roused his mother in the middle of the night. She smelled burning and made her husband open the door – to be greeted by a burst of flames. In their night gear, with the utmost difficulty and not an instant too soon, they made their way down the narrow burning staircase, Gilbert holding his son, his wife, poor woman, clasping a bedroom candlestick instead of the gold watch she imagined she was saving. I remember my father making (to my mother's great alarm) the most daring attempt to rescue belongings from the burning house, and it was only natural that armfuls of botanical specimens were what he succeeded in carrying forth.

All else was lost in that terrible night: home and shop and their total contents; Baker's botanical library, notes and collections; the entire stock of his recently-published *Flora of North Yorkshire*; the herbaria of John Storey and the elder James Backhouse, which he had on loan in that connection; much of Watson's material of *Rosa*, a genus in which Baker was then specialising; all the Reports and records of the Thirsk NHS; and many thousands of duplicates of rare and critical British plants that belonged to the Club.[19,20]

Fortunately that year's Distribution had already taken place and his chemist friend and neighbour, William Foggitt, quickly stepped in with replacements for the duplicates from his own large collection.[4] The Society's Secretary, too, was able to reconstruct its membership and finances from the answers to a circular. But otherwise the damage was irreparable; and insurance covered scarcely a tenth of this.

Baker's fellow botanists rallied round magnificently. Contributions were invited to enable him to replace his books, and a handsome amount for this was soon forthcoming. Foggitt stood in for him as Curator and relieved him for the while of that worry; and William Borrer invited him down for a stay at his home in Sussex on the pretext of needing his help in the arranging of his cryptogams. It was while engaged in this task, some eighteen months later, that he was sounded out about a possible opening at Kew, by J. D. Hooker, its new Director. "I expect soon to be in need of a person of *careful*, *neat*, *accurate* and *industrious* habits, who has made some progress in Systematic Botany and is really fond of the pursuit," Hooker wrote.[21] Baker was all of these; and inside a month the Kew Herbarium had acquired him as First Assistant.

This meant, though, that he now had to move to the South. What was to become of the Club, and for that matter the Thirsk Natural History Society, in the absence of the one who all along had been their mainstay? Foggitt, for all his goodwill, was not in a position to carry on the Distributions alone and there were no others locally who suggested themselves in his stead. Accordingly, in that January of 1866 Baker broke the news to Watson that the Club would have to be dissolved unless the London botanists could make some arrangement for continuing it themselves. He himself was willing to give some of his spare time to it but was unable to offer more.[22]

Watson at first nursed the hope that with Baker's arrival on the scene Kew might be made "a general Centre of *British* Botany" and in the process provide the Club with the modest facilities it needed. This could be a means, he argued, whereby Kew could answer the impending Parliamentary criticism (fomented by him?) that it had turned into "a mere monopoly for the advantage, scientific and pecuniary, of a very small coterie".[23] But on putting the suggestion to Hooker, he immediately had it rejected: Baker could not be spared for anything outside his official duties, he was firmly assured, and no accommodation for the Club could be given. "He holds British botany in a sort of contempt," Watson resignedly concluded.[24]

Deprived of this straw, Watson reviewed the possible alternatives. He himself was "getting past such work from age and other ties"; Syme was otherwise committed; Newbould was willing to lend a hand but to do no more than that. Baker, for his part, had too little leisure to make it safe for him to be trammelled with the Distribution, he advised, unless there was a good certainty of help; moreover, "London is a true centre, Kew is not".[24]

Luckily for posterity, Baker persisted. The help Watson counselled him to seek was in due course found – in the person of the genial and energetic Trimen, who by that time had started in medical practice and was resident in Bloomsbury. He and Baker accordingly now became Joint Curators, and the new title was assumed of the *London* Botanical Exchange Club.

Thereupon the Thirsk NHS expired. Its botanists alone, it would seem, had effectively been sustaining it and, without the impetus which the Club had provided by its presence, there simply remained too few of these with sufficient determination to continue. Although the members of the Club had nominally included a dozen or more with Thirsk addresses, only one or two of these have specimens to their names in any surviving herbaria, from which one must infer that their appearance in the lists had largely been window-dressing, in an effort to disguise the fact that the host body had embarrassingly little substance botanically. As if to confirm this, as soon as the Club moved South almost all of its Yorkshire members dropped away (including even stalwart Foggitt), never to rejoin. For all that any outsider could tell now, the Botanical Exchange Club had never extended into the North-east at all, let alone once had its home there. The years at Thirsk had started as an accident; they had ended by appearing freakish.

CHAPTER 7

The Years of Obscurity

On its emergence in its new London guise, in 1866, the Club was down to just twenty-three members. This proved but a temporary slimming, though, for within a year fourteen more had been added, as some of the previous adherents caught up and fresh recruits came in.

But the promise of a settled existence once more quickly proved short-lived. In 1868 Syme at last inherited his Scottish estate and prepared to abandon London for permanent exile in Fife.[1] By that time his formidable undertaking, a new and updated edition of Smith and Sowerby's *English Botany*, had been reaching the botanical public, volume by volume for a full five years. Critical in its taxonomic approach, employing a narrow interpretation of species that stood in sharp contrast to George Bentham's contemporaneous *Handbook of the British Flora*, that bible of the 'lumpers' (that time-honoured term to describe those who favoured a very broad interpretation), this could hardly be carried on at the same high standard with only Syme's limited personal herbarium for reference. Accordingly, to ensure that all material sent through the Club, especially the novelties, now came before him,[2] he offered to relieve Baker and Trimen of the chore of the annual Distributions. Such a step was in any case fitting, for, as he readily acknowledged, almost all his distributional data were drawn from Watson's *Cybele Britannica*, so that to this extent his work was substantially the product, at one remove, of the Club and its BSL predecessor.

Baker and Trimen having both become caught up in the editing of the *Journal of Botany* by then, they accepted this offer eagerly. Accordingly a switch-round of offices was arranged and, reverting to the pattern of the old Society, Syme found himself designated the Curator once again with two Secretaries (into which the other two were now transformed) to carry out the more purely administrative duties. The next year, in recognition that the title was in consequence no longer appropriate geographically, 'London' was quietly dropped and as the 'Botanical Exchange Club', pure and simple, it was henceforward to remain and see the rest of the century out.

The division of functions soon gave rise to trouble. Although the Curator was responsible for the writing of the annual Reports and the authorship of these was officially credited to him alone, they passed through the hands of the Secretaries on their journey to the printer and Trimen, for one, could not resist the temptation to insert a well-meant footnote or two of his own. The third year he did this, 1871, he had the ill luck to make Watson seem to be contradicting himself – and to leave the offending footnote for once anonymous. Outraged by what he called this "petty subterfuge", Watson devoted a whole appendix of his next publication to ventilating the matter, with his usual disproportionate fierceness, concluding with the announcement that as a result his contributions to the Club must cease thenceforward.[3] Already sore with Trimen for having reviewed his latest work in what he considered unfair terms, he had been spoiling for a fight and this further issue was no doubt merely a pretext.

It was a sadly unworthy ending to an association which had proved so fruitful for British botany over more than thirty years. Without Watson's hard-headedness, his forceful steering and his quickness to intercede at critical junctures there might well have been no Club for him to make this final, fist-shaking gesture against. Certainly without him these annual exchanges would never have acquired a wider scientific purpose and would have remained as they had started out, as a mere mart for narrow-minded private collecting. Perhaps indeed it was his very consciousness of all that the members owed to him that led him to behave like an emotionally blackmailing parent. If a pleading for forgiveness was what he expected, however, it was not to be forthcoming. He was allowed to go – and he never did come back. Although his name fleetingly reappeared in the Report for 1881 in the list of those who had contributed, that was the year in which he died and it is more probable that the specimens concerned were the product of a clear-out by his executor than of a death-bed reconciliation.

With Watson safely out of the way Babington, his foe of many years, pointedly now joined at last. He was, though, only one of many. For two years earlier, noting an ominous build-up in the proportion of the membership failing to contribute, Syme had launched a special effort to bring within the Club more of the subject's abler workers. At the same time he had introduced a regulation making it a condition of membership henceforward that an annual parcel be sent in. The combined result of these measures was a decided strengthening of the Club in both quality and quantity. In 1871, numbers jumped abruptly from 48 to 65, and among those who joined were such leading figures as J. E. Bagnall, subsequently the author of two county Floras, and six younger men, all in their twenties (W. H. Beeby, F. J. Hanbury, F. Arnold Lees, the clergymen brothers E. F. and W. R. Linton and

F. Buchanan White), already becoming specialists in particular, difficult groups of plants and also to write various local Floras. Never again would a single year produce a haul of such richness.

It was all the more ironical, therefore, that this sudden upwards lurch was followed almost at once by a crisis that held the very real threat of extinction once again.

The cause, as before, was a fatal hiatus at the centre: Syme was taken ill and for the greater part of 1873 was unable to perform his duties. As the specimens for distribution had all been sent to him in Scotland, it was impractical to rectify the situation from London. By that time, in any case, Trimen had bowed out – on succeeding officially to the *Journal of Botany* editorship on Seemann's death in 1871 – and Baker, while content to soldier on as the now sole Secretary, was weighed down by the calls of his career and his role had become little more than nominal.[4]

His illness apart, Syme was tired of the burden in any case. He was over sixty now, in doubtful health and effectively isolated, unable to afford to travel. For he was very hard up: his estate was a small one and heavily mortgaged, yielding an income that observers put at scarcely more than Watson's; the house was in a dilapidated condition and the garden a jungle. Cut off from fellow botanists, he was beginning to lose his former studious habits.[5] Moreover, the last of the volumes of *English Botany* that were for the time being publishable had by now issued from the press, bringing to the author an honorary doctorate, but to the publisher a heavy loss.[6] It was an obvious time to shed, at least in large part, so demanding a commitment.

At this point, providentially, an experienced youngish botanist revealed that he was planning to settle in Edinburgh and, on being sounded out, expressed willingness to share the work of the Curatorship with Syme, who was located so conveniently close. This was J. F. Duthie, a sometime pupil of Babington who, unable to follow a profession through lung trouble, had been living (and botanising extensively) in Malta. An aunt of his had belonged to the Club for several years and only recently he had won some glory by discovering a rare Milkwort, *Polygala amarella*, in Kent. He must have seemed the perfect find.

As he was not expecting to take up residence in Scotland till the middle of the summer of 1874, it was decided to suspend the Club's operations for a year and combine two Distributions in one. The publication of the Reports was thus pushed so far into arrear that a single one had to do duty for 1872–74 together.

The worst then happened. No sooner had the new Joint Curator been announced than, contrary to all expectation, he landed a job: the Professorship of Natural History at the Royal Agricultural College, Cirencester (a post which had already succoured two BSL predecessors of his). Only a year after that he left Britain altogether, to take up an appointment in India.

The Club's run of ill luck with volunteer Curators was now just as bad as that of the old London Society with its paid ones. Although it was no longer a task so impossibly laborious that no one could be found to undertake it without the inducement of money or preferential helpings of specimens, it was assumed to be beyond the capacity of anyone engaged full-time in an occupation. Unfortunately the field botany élite at this period had no men in its ranks with the requisite freedom from other demands on their time who also had the necessary ample energy. The only apparent candidates were all semi-invalids. Despite having had such slight success with one of these already, the Club could see no alternative but to have recourse to this hazardous market again.

The next such person to be found was T. R. Archer Briggs, then preparing his excellent *Flora of Plymouth* and a member of the Club since its Thirsk days. In view of his remoteness in the far South-west it was felt advisable to appoint in addition two 'Local Secretaries' (a further harking back to the BSL). One of these was Syme, the other the Manchester botanist Charles Bailey. Their role was left unexplained, but doubtless it was to share the load in one way or another and generally to be available on the sidelines in the event of faltering on the part of the Curator.

All went well for another three years. The Report for 1876 was a bumper one, indeed at forty pages it was double the length of those of a year or two earlier. But that initial surge proved too great to be sustained and in 1878 Briggs, in turn, intimated his wish to resign – though, like Syme, he was happy to stay on to the extent of taking a back seat as an additional Local Secretary. The Report for that year was edited by the ever-enduring Baker, but in it the appointment of a new Curator, in the person of A. R. Pryor, was meanwhile announced. Pryor was then working intensively on his forthcoming *Flora of Hertfordshire* and had few equals for field prowess, but just like his two predecessors he had a record of indifferent health. No one therefore should have been too surprised when he had to give up almost as soon as he had started. A severe bout of illness in the following winter forced him overseas in search of a better climate, and only a year after that he died, at the early age of forty-one.

Thoroughly demoralised by now, the Club was finally forced to face the fact that it might have to dissolve itself. In March 1879 the *Journal of Botany* reported, regretfully, that this was a possibility only too real, unless another botanist of competence could be persuaded to come forward. Never before had the future of the Distributions been more uncertain, never again was the continuance of the Club to be more precarious.

At the eleventh hour, a month or two later, a saviour appeared. And from a quarter that no one can have been expecting: he was not only a man in the best of health, but he was also fully stretched in a business career and had private commitments as well which alone most people would have supposed heavy enough. What was in it for him, some of the members may well have asked, that he should be willing to take on this further and very tedious task? It cannot have reduced their suspicions that the address to which they were invited to send their subscriptions and their parcels was that of his employers, Ralli Brothers, a well-known firm of East India merchants. Had the Club surrendered itself to Trade? Was a mere businessman attempting to buy himself into the top drawer of British botany?

But Charles Bailey was a remarkable person. Thorough, unfailingly reliable, even-tempered, supremely methodical and punctual, he was not merely a born bookkeeper but also a compulsive one. Already by this time numerous societies had appointed him their treasurer and one more to care for, he probably decided, would hardly make much difference. Keen on botany from his twenties, he was now just past forty and engaged in building up what was eventually to be the largest private herbarium in the country – so large indeed that when he came to move out of Manchester and go to live near Blackpool he found it necessary to purchase two houses, one for himself and his wife and the other next door to accommodate his collection. As with so many other Victorian men who had risen to wealth in the industrial cities, the collecting which formed the normal entrée in those days into all such pursuits had gradually turned into an end in itself. Increasingly he no longer collected plants himself but concentrated on buying up other people's collections. Indeed for a man of means he travelled peculiarly little, throughout his years with the Club becoming personally known to hardly any of the members (which at least made for an obvious choice when the time came to make a parting gift to him: an album made up of signed photographs of each and every one of them). Even his successor, after several turns as Distributor, only met him for the first time when on the point of taking over: he was, he at last discovered, "a tall, good-looking man of easy address, with kindliness beaming on his face".[7]

On agreeing to serve, Bailey found himself in much the same position as Watson in 1842: so desperately wanted that he was able to dictate his terms. What these were was soon revealed in a circular to the members. He was offering to undertake the Secretaryship – or, as a subsequent note by Baker put it, in what may have struck some as ominous business parlance, the "general management" – provided the Club was reconstituted on a narrower basis. He would discharge all of the administrative duties, including the receipt of members' parcels, but instead of a permanent Curator to handle the scientific side and make up the return parcels there would be a Distributor appointed from among the members in turn on an annual basis only. To make the work of this Distributor more manageable, it was proposed to prune the numbers considerably, at least for the time being restricting membership just to those who had contributed to the last Distribution.[8] About thirty people, almost all of them minor figures, were shed as a result.

On this obviously more workable pattern the Club now settled down, enjoying a much-needed period of stability under Bailey's quietly efficient oversight for what was to be in the end the astonishing span of twenty-four years.

The only serious upset during this entire time occurred right at the very start. George Nicholson, of Kew, it had been announced in that opening circular, had volunteered to act as the initial Distributor; but evidently he had to withdraw and, in default of anyone else, Bailey decided to do the job himself just that once.[9] This provoked a "by no means pleasant" circular from the hands of E. F. Linton and one or two others, objecting to the departure from the well-established principle that the scientific and administrative sides of the Club's affairs were conducted strictly independently.[10] Why such strong feelings should have been aroused on this at first sight trivial point can only be guessed at. Possibly the earlier brush between Watson and Trimen had alerted those who were constitutionally minded to the acrimony that could result through not keeping the two spheres divorced. The Distributor, by implication, was someone of acknowledged botanical standing and by that measure someone acceptable to pronounce, if need be adversely, on the specimens sent in by his scientific equals (or inferiors); the Secretary, on the other hand, had no necessary qualifications other than the ability to keep things running smoothly. Another possibility is that the move was more directly *ad hominem*. Linton, a grandly-connected 'squarson' (Wilberforce's term for squire-cum-parson) who made rather a point of high moral rectitude, was perhaps the member of the Club most likely to be averse to the commercial colouring of Bailey. The two may indeed have met while Linton was briefly a Manchester rector some two or three years earlier and may perhaps have had their differences. What was certainly true, though, was that Bailey, unlike for example Linton, had no acknowledged standing as a critical botanist;[11] his action may thus have had the appearance of pretentiousness and given rise to resentment as a result. Whatever the real reason for the outcry, the complaint was pursued relentlessly. The following year's Distributor, James Groves, was required to write to all the members and ask them for their opinion on the matter; and when the majority, perhaps predictably, agreed that the Secretary should not usurp the role of the Distributor and make critical comments on the specimens submitted, it was made clear in no uncertain terms that their verdict was to be regarded as binding. A lesser man would have responded to this by sending in his resignation; the long-suffering Bailey, however, was content to acquiesce.

The incident had an important repercussion. Although Linton had won his point, relations were inevitably left somewhat strained and for the next few years, while not going so far as to resign, he preferred to transfer his energies elsewhere. Consequently when, in December 1884, a group of obscure young Quakers banded themselves together as an alternative body under the title of the Watson Botanical Exchange Club, Linton's name, together with that of his brother and his father-in-law, J. D. Gray, featured, conspicuously and incongruously, in the list of founder members. This weighty initial patronage from Linton and his family at once gave the new club a solidity and respectability that it would otherwise have lacked and so helped to ensure its survival.[12] A year or two later the Lintons made use of it in turn to publicise a British branch which they had gone on to establish of the Linnaea Association, a new international body designed to link together the various exchange clubs that had come into being in different countries and to expand their activities into a world-wide traffic.

For by that time the long-continued success of first the BSL and then the BEC had attracted numerous imitators. In 1853 a Phytological Club formed in connection with the Pharmaceutical Society included the exchange of specimens among its functions, as did the later Society of Amateur Botanists. In 1858 the Geologists' Association, even though abortively, contemplated starting a similar scheme for fossils; the next year, equally abortively, there was talk of a Co-operative Entomological Society. In 1878 a short-lived exchange club was launched under the auspices of the popular magazine *Science Gossip*, and another periodical, the *Entomologist's Record*, was later to copy its example in 1890. Later still a Moss Exchange Club was to be founded and some while after that another one for lichens.

Unlike most of these other initiatives, the Watson Club lasted – and for no less than forty years. The reason was not merely that it was run efficiently, with consistent dedication: it also filled a definite gap. In truth the two Clubs were not so much rivals as usefully complementary. The senior body catered for the more sophisticated and experienced, the newcomer specialised in bringing on beginners. But although they had at first very few members in common, over the years they overlapped increasingly and from time to time people drew what seemed to them the logical conclusion and urged that the two be amalgamated. Such proposals, however, were misconceived. The hard reality was that an exchange club run on voluntary labour had a maximum viable size: beyond more than twenty contributors, a later Secretary reckoned, the number of parcels to be dealt with became a serious infliction.[13] Allowing for a small proportion whose requirements were of the slightest, both clubs consequently kept the total of contributors down to twenty-five or so throughout much of this period. Admittedly when the Moss Exchange Club was founded, the Secretary of that called for a halt only on reaching thirty-six[14] – but then bryologists keep their specimens in comparatively small packets, which must have made for much easier handling.

It was not just the time and effort involved in making up and sending off the parcels that Distributors found so formidable. There was the problem also of space: the gatherings had to be laid out in rows in order to take from each in turn the necessary sample specimens. Bailey had the advantage of a one-time stable-cum-coach-house available for this process,[15] but most were less fortunate. Altogether there were about seven processes to go through, one Distributor estimated, before the stage was reached of compiling the Report.[16] Another Distributor, J. W. White, found the sheer mechanical work so unexpectedly prolonged that he had little time left over for critical examination of the plants sent in and so had to apologise shamefacedly for his eventual "rather bald" write-up.[17] In order to prevent this kind of miscalculation recurring Bailey, with characteristic thoughtfulness, made it an invariable practice to offer to relieve the Distributor of the entire manual side of the office.[18] But not all Distributors had the good sense to take him up on this.

A further bugbear of Distributors was the mass of bad material sent in. "Some of the parcels are, as usual, simply shameful," one of them was moved to complain bitterly; "it is abominably selfish to . . . send such worse than useless rubbish."[19] Charles Waterfall, of Hull, was a particularly notorious offender. Year after year, with a blithe unconcern, he poured forth torrents of specimens that were invariably appalling: carelessly dried, wrongly named, illegibly labelled – and, worse still, they were dockyard casuals that nobody really wanted. Repeated remonstrances had absolutely no effect. In the end, after a widely-supported proposal that his next parcel be refused and returned to him, he was pressured into resigning[20] – and was promptly admitted to membership of the other club instead. A juster fate would have exposed him to the ruthlessness of Linton, under whose Distributorship in 1890 no fewer than sixty-seven gatherings were consigned to the newly-created category 'Destroyed as useless'.[21]

Those who did send passable material, though, were liable to offend in their turn by sending in too little. The regulations laid down that each gathering had to consist, as far as possible, of enough specimens to fill a minimum of ten herbarium sheets, and for preference twenty or even thirty of these. Unless this requirement was observed, insufficient were left over after the needs of the national herbaria and the few largest contributors had been met.

Naturally this prescription as to quantity was intended to apply only to taxa of critical interest, a category into which by this time most of the material submitted came (in 1891, for instance, half of all the gatherings were of brambles). In the case of well-marked species sent in merely as vouchers for new county records members were assured that a single specimen was sufficient. But inevitably these conventions were on occasion abused, and in any case even ten specimens removed of some critical species (in particular the many very rare and too easily uprooted hawkweeds) could mean that a population was imperilled or extinguished. Now that more and more people were developing a conscience in such matters, the exchange clubs were rightly condemned for the heedlessness of their practices. But too often they were the targets for shafts of sheer emotionalism, attacked merely as Collecting's institutionalised embodiment, denounced as nothing more nor less than a series of extermination agencies. Such blanket condemnation was absurd. Damage on the scale they were frequently accused of had indeed occurred, but in more innocent days half a century earlier: the harm they did now was a mainly marginal by-product of an enthusiasm that had grown much more genuinely scientific.

The high average level of taxonomic sophistication to be found now at least in the BEC had had the effect indeed of making the members more and more choosy as to the kinds of plants they wished to receive. As a result increasing emphasis was placed on the Desiderata List, the growing length and complexity of which imposed a considerable extra strain on the Club financially. At the same time the longer, much more meaty Reports that had now become standard added to the cost of printing. To meet these heavier expenses, the subscription ought to have been raised from its modest five shillings or at least the disgracefully numerous members who had fallen into arrear should have been treated less leniently. Alternatively, or in addition, the members could have been asked to meet the postage on the parcels they were sent: it was only their own incoming ones that travelled carriage-paid. But Bailey was too kind-hearted, or else feared to provoke the loss of several

Revd E. S. Marshall.

of the most valuable contributors; Linton, for example, was one of the largest of the defaulters but would surely have resigned if pressed (at any rate by Bailey). Instead, Bailey preferred to make up the annual deficit out of his own pocket, doubtless regarding the sums involved as too trivial, by his own personal yardstick, to be worth the pain and effort of collecting them by the proper courses. Altogether, over the years, these secret subsidies to the Club amounted to over £200.[22] Though Bailey would have liked to think of them as charitable donations and could readily have cited cases where private generosity had similarly buoyed up other bodies[23] (most notably the Entomological Society, by J. W. Dunning), he had all the same set a most undesirable precedent. Apart from the fact that it is wrong in principle for any organisation to have its true financial position concealed from it, there was bound to come a time when the subsidising would have to stop and the adjustment then would be all the sharper and more traumatic for having been so long deferred. It was unfair to leave such a legacy to a successor, who was bound to be embarrassed, if not a man of ample means, by inability to prolong the indulgence.

Eventually, even Bailey wearied. There was so little thanks, there were so many petty complaints in return for so much expenditure of effort. "I am reluctantly coming to the conclusion," he confided to a fellow member in 1898, "that a business man like myself is unsuitable to hold the office of secretary of a botanical exchange club ... I think the time is coming when it must pass into other hands."[24] And probably he was right, for he was beginning to acquire a reputation for not wanting to be bothered with queries, for getting things done late.[25] Even so it was not until 1902 that he finally made up his mind to go. And even then it was only the acquisition of further heavy private commitments, demanding all his energy, that propelled him into action.

His impending departure at once raised the question whether the Club could, or even should, continue. Apart from its persistent insolvency and the obvious difficulty of finding another paragon like Bailey to run it, there had lately been murmurings that the Club would be better replaced by some body much broader in scope. The study of the British flora had, after all, grown up a lot since the Club had started its existence and there was now a much wider botanical public whose interests went beyond the niceties of critical taxonomy and even excluded collecting altogether. Though it was too much to hope that the new world of professional biology which had emerged since the 1860s would develop lines of enquiry that would lead it to want to share the same organisation as the main mass of amateurs, the first glimmerings were on the horizon of a possible unifying middle way in the form of the study of plant communities.

As the Victorian era drew to a close there was a widely-shared feeling of staleness in many branches of learning, a conviction that everything it was possible to establish by means of the traditional approaches had now been established to the uttermost. There was a general groping for a change of direction. Thus the retirement of Bailey came at a time when stimulating new vistas were beginning to open up. As well as marking the end of an era for the Club, it usefully coincided with a natural break in botany more generally.

The final quarter of the century had been a time for stocktaking, for grand summations of all those years of tireless industry, and in their own special way Britain's field botanists had had their share of this tendency. For the critical taxonomists among them the obvious step had been the producing of a new national Flora. Syme's *English Botany* was on too lavish a scale to serve as a workaday tool; Babington's *Manual*, the best of the portables, had not been revised for almost twenty years; 'Hooker's' *Student's Flora* (in fact the work of Baker) was too compressed and technical; Bentham's *Handbook* was essentially for the non-critical. The gap was a glaring one, and for some time Linton had been known to be accumulating material with the intention of filling it. In 1898 he eventually allowed the *Journal of Botany* to come out with an announcement[30] (an unwanted rival having appeared on the scene)[31] and as a result A. R. Waller, the founder of the Watson Club, approached him to suggest the firm he worked for, Duckworth & Co, as the publisher. Nothing more, however, was heard of the venture.

The devotees of plant geography were luckier. In 1883 an enlarged, posthumous second edition was produced of Watson's culminating monograph, *Topographical Botany*. In both this and its predecessor the distribution of the flowering plants and ferns of Great Britain was set out in terms of a mapping unit invented by him. This was based on the boundaries of the administrative counties, but by amalgamation of the smaller ones and splitting up the larger ones he divided mainland Britain and the more immediately offshore islands into 112 units of approximately equal area, which he termed 'vice-counties'. The evidence for the occurrences of plants in each of these was meticulously referenced to published or unpublished lists or to specimens in his own herbarium. It was a massive and dependable foundation on which all further recording in Britain would be based for a full three quarters of the century that followed.

The equivalent covering of Ireland inevitably lagged behind. A. G. More, the most patiently dedicated of Watson's followers, had toiled for years to bring up to date the attempt at a *Cybele Hibernica* which he and David Moore had produced in 1866 and by the time of his death, in 1895, that task was very near completion. Three years later it duly found embodiment in print, seen through the press by Nathaniel Colgan and R. W. Scully, themselves the respective future authors of two of the best of the Irish county Floras. But by that time the coming king of Irish botany, R. Lloyd Praeger, had embarked on a project designed to jump a generation farther on. Though this required a huge amount of field work, he was able to carry it through to publication in 1901 in the space of a mere six years. For the purpose he divided Ireland into forty vice-counties and proclaimed in his choice of title for his book, *Irish Topographical Botany*, that it was intended as the counterpart of Watson's. All were at once agreed that it made a worthy companion to that on the library shelves.

So enthusiastic had been the reception of Watson's long-awaited work that a Botanical Locality Record Club had sprung into being immediately on the appearance of the first edition in 1873.[26] Founded at the instance of F. Arnold Lees, the surgeon author of the *Flora of West Yorkshire*, and T. B. Blow, a sometime professional beekeeper, this had as its declared aim the setting up of a national network of recorders, "just three or four botanists in each county", who would maintain an up-to-date tally of the plants occurring in their local areas and send in voucher specimens in support of new records. In essence it thus performed the role of the present-day Records Committee of the BSBI and was usefully complementary to the two exchange clubs in its turn. In due course the omnipresent Bailey agreed to act as its Treasurer as well and thenceforward its links with the BEC were strong. A high standard of accuracy was ensured by a strict rule that every record published in the annual reports had to be backed by a voucher specimen, the material received as a result being donated at the end of each year to the Herbarium at Kew.[27] After some years of very useful work, however, the Record Club eventually petered out, around 1890, when Lees, its main driving-force throughout, became heavily preoccupied with his professional duties.[28]

But by then the coexistence of all these three bodies with memberships and functions that substantially overlapped had powerfully reinforced the case of those who yearned for some comprehensive grouping. In 1890 F. H. Ward, a South London medical officer and microscopist who was at that time the foremost contributor to the Watson Club, accordingly went so far as to try out ideas along these lines on Linton:[29]

> I have been thinking over lately the possibility of forming a new botanical society for it seems to me there is not one at present that does much to encourage that study in England. My view of the matter is that

> what is wanted is a society meeting once a month in London where papers might be read and specimens exhibited and all matters relating to the subject discussed, that there should be an exchange department, and a journal published periodically in which the papers or abstracts of them should be published as well as records of new plants or new varieties or new stations of the better-known ones. I think the aim should be to do the work of the different exchange clubs [and] the record club, as well as popularising and extending the subject itself.

The Linnean Society, Ward went on, was too expensive and in any case did not meet the needs that he had mentioned. Nor was it likely to be enough for the three existing clubs simply to be merged and the resulting single subscription to be fixed at double their current levels (as was already being canvassed as an alternative), for this would fail to yield a sufficiently large income to allow anything more than an amalgamation of the reports. "If there should be a Society of British Botanists" – and here an approach to the future title surfaces for the first time – "the only claim to enter which being the payment of the subscription," then he supposed there would no doubt be members having no claim to be termed botanists. This, however, he would by no means object to, as long as they kept up their subscriptions.

Although nothing came of these ideas just then, their general theme was one that was sure to re-emerge. The only question was whether the expansion, when it came, would take place within the framework of one or both of the surviving clubs or whether there would be created a new organisation altogether.

CHAPTER 8

Civil War

In view of the cardinal importance of the decisions with which the Club was faced, Bailey had felt it only appropriate to go to the exceptional length of a postal ballot of the members. To this end a questionnaire was drawn up and sent out in September 1903 to the forty-two who were nominally on the books.

All but six replied and their vote overwhelmingly was in favour of the Club's continuance. The possibility of a merger with the Watson Club was not canvassed formally, but seven volunteered that as a solution instead (though Bailey himself was strongly against). The great majority proved willing to accept the raising of the subscription to half as much again.[1]

The third question posed was whom they wished to see in Bailey's place in the event of a decision that the Club had a future. One member, the much-respected Revd Augustin Ley, responded to this by proposing Baker; but as the latter, by then on the threshold of seventy, saw this as a fitting juncture at which to retire from his half a century's association with the Club, this turned out not to be a viable suggestion. It may have been intended, though, to insult by its very absurdity: for the only serious candidate, as everyone knew – but a weighty section of the membership secretly abhorred – was the Oxford shopkeeper, George Claridge Druce. In the event, of the alternative names put forward, his was the only one that attracted any substantial measure of support; and he was accordingly declared elected.

From some points of view the choice of Druce could hardly have been bettered. He was prodigiously energetic; he was about to have abundant leisure (for within eighteen months he would be retiring from business); he was an excellent field botanist of wide experience and noteworthy *élan*, with already three county Floras to his credit. Certainly, too, he had earned the office: an active member since 1878, he had served as Distributor on as many as four occasions and for long had furthered the Club's interests enthusiastically and devotedly. In the eyes of Bailey at any rate he clearly seemed a thoroughly worthy successor.

Yet much of the field botany Establishment had come to look at him askance. Its objection is best summed up in these words of Revd E. S. Marshall, in a letter written around this time: "... He is so 'pushful' and so hasty in his work – the result of an imperfect education, for his mind is quick. He also puts too many goods into the shop-window."[2] Another contemporary was once characterised by Marshall as "inclined to be rather 'cocky' – like a second-rate Druce, with far less experience and cleverness".[3] Many of the clerical members found his driving egotism hard to take, and it may have been no coincidence that their number slumped sharply as soon as he succeeded to the Secretaryship.[4] But matters were made worse by Druce's inordinate sensitiveness to criticism, the mildest suggestion of which he was liable to take as a personal affront. As it was his botanical equals who mainly found cause to criticise him, so it was those with whom he was continually falling out, resulting in estrangements that in some cases endured for years. "For a person of pleasant demeanour," a friend was moved to exclaim, shortly before Druce assumed the Secretaryship, "it seems to me you are always quarrelling!"[5]

Much of the explanation for this touchiness, as also for his insatiable need to be admired, lay in his difficult early life. As we learn from the *Dictionary of National Biography*, he was born illegitimate and brought up in straitened circumstances in a village in Northamptonshire, as an only child without "the advantage of frictional companionship" (in the words of one of the most perceptive of his obituarists).[6] His brightness was spotted early, however, and receiving a better education than might have been expected he was enabled to qualify as a pharmacist. After some years in Northampton working for the firm of Jeyes (proprietors of the well-known disinfectant) he moved to Oxford and set up in business on his own behalf, his shop in the High Street there gradually acquiring renown among generations of undergraduates – not least for Druce's 'purple specials', the favoured nostrum throughout the University for the curing of a hangover. By 1886, when he made his botanical reputation with his *Flora of Oxfordshire*, he was already in sufficiently high standing locally to have an honorary MA conferred on him for this. Not long after, he entered local politics and was in due course elected Sheriff of the City and then, in 1900, Mayor. An ardent freemason and a keen philatelist as well, the numerous friends and contacts these multifarious activities brought him helped him to prosper greatly in his trade. Unmarried, spending very little except on his fine botanical library and extensive stamp collection, an adventurous investor (to judge from his dividend counterfoils) and a moneylender in a small way on the side, he ended up an extremely wealthy man.

If the fact that he was an affluent man of business had made Bailey suspect in the eyes of some of the Club's more influential members, there was all the more reason for Druce to be regarded warily on this account too. For, unlike Bailey, he had been his own master for most of his working life, unaccustomed to working in a team and able to do more or less as he wanted. Unlike Bailey, moreover, he was experienced in the ways of politics and socially ambitious. Free from any constraints of time or money, he was the very last person who should have been entrusted with the affairs of a voluntary body without a tight constitutional harness. In justice to the members, though, it must be stressed that it was a mere exchange club that he had been asked to take over; there was no reason to suppose that this would not be continuing along the lines on which it had long been run, and that therefore no call was seen for any set of rules over and above those which governed the terms and mechanics of the annual Distributions.

The very first Report to issue from him as Secretary carried ominous signs of what was to come. For a start, the whole of it was contributed by him alone – admittedly fortuitously, for he had been asked to act as that year's Distributor before anyone had known that he would be succeeding to the Secretaryship, and his botanical standing was such that his competence to act in both roles could hardly be challenged as it had been in the case of Bailey. Instead of confining himself, however, to the mere one or two paragraphs of introduction that was the most his austerer predecessors had seen fit to rise to, he produced a six-page editorial shading off into a readable review of some of the chief finds of the year, not necessarily all of them brought to notice through the Club. Appreciated though this was, it looked worryingly costly, even given the big rise in subscription that had now taken place. But more worrying still, at any rate for those who wished the Club itself to remain unchanged, must have been one particular sentence that Druce dropped in. Although he recognised, he said, that many members felt it would be inadvisable to lower the standard in any way, "at the same time there appears to be in our own Society every facility for the less expert botanist who may be admitted to our ranks". How many who read those words, one wonders, realised that the use of 'Society' was no mere slip of the pen but a start at preparing them, as if subliminally, for a drastic change in the whole character of the Club?

In the next five years most of Druce's efforts on behalf of the Club went into building up its numbers. Every botanist in the country who was conceivably recruitable seems to have been approached by him: not only the many amateur collectors who for one reason or another had never got around to joining, but also the now growing number of university professionals. Former members, too, were pressed to return. Even Linton, who had chosen to resign at last when Druce became Secretary, was assured, in a rather double-edged compliment, that the Club "seems like *Hamlet* without the ghost in your absence".[7]

With the membership fast nearing the hundred mark, thus demonstrating that much growth potential did indeed exist, Druce decided the time had come to reveal his plans more openly. This he chose to do in the form of a 'Note to members' printed as a tailpiece to the 1908 Report. It read as follows:

> It appears to me that in some ways the Club might be made more useful if its bounds were widened and its activities extended so as to make it a Society of Field Botanists as well as an Exchange Club. The proposed Society might be organised on the plan of the old Botanical Society of London, which did such excellent work for many years. There can be little doubt that Systematic Botany in this country is, in certain directions, languishing through the want of some central organization.
>
> At present, many really good botanists hesitate to join us, some because they think it wrong to collect large quantities of specimens, and others because they think exchange clubs lead to the extirpation of rare plants. These are, to a large degree, mistaken views, but they obtain ... It is highly probable that many of the botanists in question would join a Society whose activities were many-sided, even though they cannot be persuaded to join an Exchange Club. With an enlarged membership, British systematists would be kept *in touch* with one another, and would be able to illustrate more fully and to describe in greater detail the results of their labours in the field. Critical plants would be more widely studied, and botanists only partially interested in the subject would be made keener. Comparative culture of critical forms would be stimulated ... Academic systematists would doubtless be pleased to join the new organization, and to contribute to a knowledge of the ecology, physiology, structure and development of critical genera. Such work is now being done, and field-botanists would profit by being in touch with such workers.
>
> A yearly meeting might be held in winter and a joint field excursion made in summer; the value of such gatherings need not be enlarged upon.
>
> I should like to know if this proposed scheme is acceptable or not to the members. The primary objects of the Club could still be prosecuted, and both contributing and non-contributing members would gain much and lose nothing by the proposed change, nor do I think the extra burden thrown upon the Distributor would be a serious difficulty. May I ask members to let me know their views on the subject.

In the event few members responded – as few people ever do when expected to take the initiative themselves instead of being sent a questionnaire – but these few were on the whole in favour.[8] Rashly, Druce decided to interpret this as sanctioning all that he had proposed; and thereupon, as a symbolic first step, he accordingly added the words, 'and Society' to the name of the Club on the next Report's title page. Two years later, emboldened, he put them on the front cover as well. No one uttered a word of protest, he was later to recall.[9]

For someone with so many years of experience in local politics this was an astonishing piece of naïvety – if naïvety it was. For he should surely have realised that the implications of what he was proposing were so sweepingly radical that nothing short of a plebiscite could suffice. Instead, he had done no more than publish what to all appearances was a mere preliminary note of consultation, pitched in the tentative terms appropriate to such a document. There had been no indication he had intended it to be taken as a manifesto, on which he proposed to act if no opposition was expressed.

Ironically, though, it was not so much the altering of the very character of the Club in this seemingly clandestine manner that aroused the wrath of a considerable section of the membership. It was, rather, the shameless way in which Druce had begun to use the Reports as a vehicle for his own personal opinions on taxonomy and nomenclature, opinions which, as he well knew, were gratingly at variance with those of British botanists in general. Nomenclature is one of those hobby-horses which are inclined to bolt with their riders and no amount of opposition was ever to be enough to deter Druce from the idiosyncratic way in which he insisted the law of priority should be interpreted. Apart from that, he was over-fond of bestowing scientific names on trivial variants, most of which he had never really studied. Both practices seemed to his contemporaries mere forms of showing-off.

One person they riled more particularly was James Britten, of the by then British Museum (Natural History), the editor at that time of the *Journal of Botany* and a man who delighted in his ability to needle people. (A Roman Catholic convert of notorious fervour, he was referred to behind his back as '*Urtica dodartii*' – the Roman Nettle.)[10] Britten too had strong views on nomenclature and it was inevitable that he and Druce would clash. Their differences dated from the mid-1890s, if not earlier, and culminated in the eventual cancelling by a furious Druce of his *Journal of Botany* subscription, a gesture he doggedly refused ever to go back upon until Britten's editorship finally ended with his death in 1924.[11] In 1898 Britten used the opportunity of a review of Druce's *Flora of Berkshire* to assail him frontally: the book, he complained, had been made "a pretext for foisting upon botanical literature a number of new names, both of species and varieties, which seem to be created mainly in order that Mr Druce may have the pleasure of putting 'mihi' after them". "The fact that in the last edition of the *London Catalogue*," he went on, "the great bulk of his published varieties were ignored might have shown him that they were regarded by some as of doubtful value."

Unable to contribute any longer to the *Journal of Botany*, Druce at first sought sanctuary in the *Annals of Scottish Natural History*. But that was too obviously regional to be patronised too regularly with non-Scottish material and for a time he toyed with the idea of launching a new British botanical journal of his own.[12] That scheme was overtaken when he succeeded to the BEC Secretaryship, for his plans for an enlargement of the scope of the Club assumed an accompanying expansion in the Reports as well and thereby carried the promise of an acceptable substitute for what he had had in mind earlier.

Had this broadening of the Reports been more of a collaborative undertaking, the volume of objections would probably have been slight. Britten had alienated numerous other potential contributors apart from Druce and, despite the risk that the *Journal of Botany*, already short of material, might be killed by further competition, the need for an alternative outlet for papers on the British flora was being felt increasingly. But Druce insisted on contributing all the material himself and he used the greatly extended editorials, now distinguished by the name of 'Secretary's Reports', to pontificate on a wide variety of matters, in much the same way as Britten had long been accustomed to doing in the *Journal of Botany*. It was this above all which created resentment. As Marshall was later to make plain to him, many members felt there was "too much pushing of your personal views and aims; too much 'booming' of your *List of British Plants* . . . too free (and sometimes acrid) criticism of fellow-botanists, etc. . . . There has been a regrettable tone of 'superiority' – probably quite unconscious – and the assumption of a right to act as a general censor of British botany – and botanists."[13] Apart from all this, members were increasingly alarmed by the cost implications of so much extra printed material. To make it worse, in the 1907 Report there was the unheard-of luxury of as many as four photographs, contributed (and doubtless paid for) by Bailey, three of them of *Oenothera* taxa growing on his local dunes. Had the subscription been increased merely to provide the Secretary with a lavishly-furnished platform, they asked themselves? And were they shortly going to be asked for more in order to keep him in the style to which he had become accustomed? Some at least of these suspicions could have been allayed, had Druce only seen fit to disclose that the cost of printing his own lengthy outpourings on nomenclature[14] had been covered by an anonymous donation (from some pocket other than his own);[15] but for some reason he chose not to.

By 1911 the annoyance of the 'old guard' was such that one or two whose friendships with Druce were of more particularly long standing decided to try remonstrating with him personally. All they received for their pains, however, were angry protestations of innocence, and even some downright abuse.[16] Clearly, some more strident and more collective move was going to be needed if the Secretary was to be made to mend his ways.

At that point, just as the opposition forces were massing, a willing champion of their cause made a timely appearance. This was Charles Edward Moss.

Moss was one of those brusque, combative men for which Yorkshire has long been renowned. The son of a Nonconformist minister, he had started out on the lowest possible rung of the academic ladder, in an elementary school as a pupil teacher; from there, by brilliance and by force of character, he had steadily made his way, via Manchester and a higher doctorate, to the ill-paid but prestigious post of Curator of the Herbarium at Cambridge. Up till then he had been one of the leading exponents in this country of the budding discipline of plant ecology; but as the course in this at Cambridge was already being taught by the even more respected A. G. Tansley, it was to taxonomy that Moss now turned his very impressive energies.

Shortly after his arrival the University Press was asked to consider publishing a set of pen-and-ink drawings of British plants made by a retired Huntingdon solicitor, E. W. Hunnybun. Consulted about these, Moss came up with a proposal for a *Cambridge British Flora* under his editorship, with Hunnybun's illustrations used, as Syme had used Sowerby's, to add popular appeal to a scientific text incorporating the results of the latest critical work. Since Linton's abortive venture of some ten years earlier, serious students had been feeling ever more keenly the lack of an up-to-date national Flora and so, persuaded that there was a lucrative gap waiting to be filled, the publishers agreed to go ahead with a multi-volume project including numerous photographs. Moss quickly enlisted the co-operation of all the country's leading field botanists and at a meeting at the British Museum (Natural History) in March 1912 outlined his plans to an enthusiastic audience of potential contributors. Marshall, for one, was deeply impressed by his obvious business aptitude as well as by his clearness of view and all-round botanical ability: "a sound, careful worker, who is likely to do much".[17]

All of a sudden, therefore, the upper echelon had acquired an acknowledged leader. Despite the recentness with which he had come to British plant taxonomy, his preceding vegetational work had secured him the requisite field standing. Moreover, he was a professional, armed with awesome qualifications: he was not only a DSc but as Curator of the Herbarium at Cambridge he was the holder of the most relevant post at the university that had the highest reputation in Britain for science. With a very strong personality, he was also entirely accustomed to taking the centre of the stage.

Initially, his relations with Druce were good. As honorary Curator of the Fielding Herbarium at Oxford, Druce was in effect his opposite number and to begin with Moss even talked of making him the Flora's joint editor.[18] As it was, Druce had been invited to write the accounts of many of the families, and his name featured among the authors in the first of the volumes when this came out in 1914. Moss, for his part, acted as the Club's Distributor for 1910 – for the Cambridge Herbarium was a BEC subscriber and so he could participate as its official representative (he never did become a member in his personal capacity).

It was in May 1915 that Moss received a letter from Marshall soliciting his aid in bringing pressure to bear on Druce. He readily agreed, but dissented from Marshall's suggested line of attack – a 'round robin' – as savouring of cowardice: instead, he proposed that about a dozen of the more prominent members should fire off individual letters of protest, the posting of these being timed so that all descended on Druce simultaneously.[19] Marshall decided to follow this suggestion and, though one or two of those he approached demurred (among them J. W. White and C. E. Salmon), preferring the less personal exposure of a collective circular, six or seven letters were duly dispatched and are known to have hit their target.[20]

In the meantime, independently, A. B. Rendle had sent a carefully worded letter on behalf of his colleagues at the British Museum (Natural History), ostensibly to complain at the bewildering number of changes in its name which the Club had recently undergone. To this Druce replied with a studied *hauteur* – and, in a neat piece of one-upmanship, on the Marquess of Bath's notepaper[21] – referring grandly to "our Society" and explaining that "we extended its scope so as [to] include many people who disliked the idea of belonging to an *Exchange Club*, but who did wish to belong to a society". Now that these were in the majority of three to one, he added, it had seemed only fitting to change the name to 'Society and' instead of 'and Society' from the 1912 Report onwards.[22] Rendle lost no time in sending this reply on to Britten, who by then was living in retirement at Brentford, and duly elicited the predictable explosion. "Surely this is the most astounding cheek! Who are 'we'?" responded Britten in fury. Marshall, too, was shown the letter and, though his reaction was restrained by comparison, he similarly found it far from satisfactory. It was Druce himself, he pointed out indignantly, "who got so many outsiders to join. He has simply swamped the regular BEC members; and now he advances this misdeed as a ground for altering the whole scope – and title – of the Club."[23]

The receipt of this volley of letters at last left Druce in no doubt at all that a concerted attempt was under way to force him to reverse his policies. The time had clearly come to put his side of things to the membership. On June 29th, accordingly, he sent out a brief circular in which he explained that "a certain amount of criticism" had arisen about the way the Club was being run and asked people to indicate – this time unambiguously, in the form of a questionnaire – whether they were satisfied with the existing management or whether they wished for a change of Secretary. He also took the opportunity to ask afresh whether an occasional meeting or excursion would have support.[24] Outrageously, he then made nonsense of this sop to democracy by enclosing individual covering letters to all his likely sympathisers, complaining that there was a "cabal" in the Club against him. Learning of this, Marshall, not unreasonably, accused him of having packed the jury.

Only two members – one of whom was Moss – had the nerve to put an outright 'no' against the first of the questions, though a good many others implied as much in their answers or in follow-up letters. None at all, however, not even Moss, wrote that they wished for a change of Secretary, though Marshall at least had the courage to slip in "but not as an autocrat". Equally, support was well-nigh unanimous for keeping both the name and the Secretary's Reports as they stood. Few felt strongly about another suggestion that the Society and the Exchange Club should have separate reports and accounts, and no one saw a case for holding a meeting of the members until the First World War was over.

If the answers to this circular appeared to let Druce off lightly, it was a very different matter with the torrent of sermonising that now descended on him in letters. How many times he must have rued that so many of the members were clergymen or schoolmasters! The veteran Linnaeus Cumming, chemistry master to generations of Rugby Schoolboys, was one of those who lectured him particularly sternly and to the point:

> In answer to your remarks I cannot regard the assent of any number of the members of the Club to the change in name and purpose as under the circumstances legalizing the change. In the first place the assent was probably given [on] a one-sided statement of the case and not after having all the pros and cons before them. In the second place at a general meeting of the Club many arguments on either side might be evoked and it is probable a decision arrived at might be an amendment on either of the two alternatives presented by letter.
> ... I feel strongly expenses should be reduced in order that the Club should not depend on your financial support. First because ... members of the Club have a proper pride and sense of independence, forbidding them willingly being in debt to any man. Secondly because you make it hard for your successor (and we dare not hope that you are immortal) if a poor man succeeds you ... I have two or three times, when speaking of BEC to botanical friends who ought to be supporters, been met with the

> rejoinder, "Oh that is Druce's club." I am afraid your financial support year by year only emphasizes such a regrettable phrase.

There was a great deal more in similar vein from his and other hands. On a man of Druce's sensitivity the pressure must have told.

But it was one thing to preach to him the error of his ways: it was quite another to shift him from his adopted course in any significant respect. If that was ever to be achieved, the membership needed to be appealed to over his head. And from that all but one of the opposition shrank, preferring to tell themselves in their faint-heartedness that there was nothing more that could be done.

The one exception was Moss. No man to shirk a fight, he decided that it was up to him alone now to try to rescue the Club from the impasse and to stop the present state of affairs from enduring permanently. Wishing to be seen to be above the fray, but at the same time anxious that the case against Druce – which as yet had been ventilated only in private – should not go unheard, he cast himself in the dual role of both mediator and prosecutor.

The broadside came on August 27th, in the form of a circular nine pages long.[25] In contrast to the elliptical, almost shamefaced style that Druce tended to favour, Moss believed in bluntness. Lucidly written and persuasively argued, the document was devastating: one by one the points at issue were rehearsed and the case against the Secretary gradually built up into an overwhelming indictment. But instead of drawing the obvious conclusion and urging that he either go back on his policies or resign, Moss leapt up abruptly into the arbitrator's chair and rejected this as a reasonable solution: "The Secretary . . . has, at least indirectly, given ample notice of his intentions: his policy has taken many years to develop; and if members have for years been irritated by his policy, and yet have lacked sufficient moral courage to make a formal protest, they should be prepared to pay some penalty for their supineness." Yet the other, equally obvious alternative, the departure from the Club of all those objecting to the Secretary's policies, he rejected as unreasonable as well. A third course of action, though, which he would otherwise have wished to put forward, had been rendered impracticable by the outbreak of the War. "It had been my intention, as some members know, to call the Club together, and ask it to convert itself into a Society with annually elected officers and with a Committee of Management. I was prepared to place before the Club a considered scheme which, if it had been accepted, would, in my judgment, have solved the difficulties. However, no one wishes to attend meetings of that kind or to form botanical societies at the present time; and I prefer therefore to postpone any consideration of my scheme until a more convenient season." He concluded therefore that there was only one possible thing to be done: to suspend the entire activities of the Club for the duration of the War. If too few members voiced their support for this, then he would have no alternative but to sever his connection with the Club.

With this last, seemingly egotistical throw Moss had completely undone his case. The one thing that most members wanted above all else was for the Club to carry on, uninterruptedly. The remedy that he proposed would be worse than the disease, protested even Marshall.[26] Moss had ended up by antagonising even some of his staunchest allies.

Worse still, though, he had given new heart to the supporters of Druce. Not until then was it apparent how deep was the loyalty to the latter of so many of the newer members. Despite all the failings of which so much had been heard already, Druce had an extremely likeable, even lovable side to his nature. To those given no cause to see him in any other light he appeared "a pleasant, gentle, friendly man, an amusing companion who enjoyed laughter".[27] "There was in him," wrote another of his obituarists, "a Peter-Pan-like boyishness and artlessness which made him a perpetual source of refreshment and delight to his friends. Wherever he went he was a vivifying and inspiring influence."[28] He was untiringly helpful to beginners and had a marvellous way with children. Instinctively, too, he felt for the underdog and those who had fallen on adversity. Only Druce would have thought of having a BEC Benevolent Fund; only he had the initiative and energy to organise a whip-round whenever a fellow leading botanist (like Alfred Fryer or Arthur Bennett) was temporarily suffering hardship. Numerous people, in various walks of life, were in his debt for kindnesses both great and small. And besides all that he had endearing idiosyncrasies: his invariable bowler hat, for example, his peculiar high-pitched laugh and his bachelor's unkemptness.

This was a very different Druce from the man whom Britten and then Marshall and now Moss had found so exasperating and slippery. This Druce was lily-white, even almost saintly in the eyes of his friends and admirers. Entirely unaware of all that had been going on behind the scenes and in higher botanical circles, they found it utterly impossible to conceive that the lengthy dossier that Moss had put around could have sufficient substance to warrant its being taken seriously. In their eyes it could only have been the product of uncalled-for malevolence.

Out of the hive they all now swarmed with an infuriated buzzing, impatient to discharge their stings in the defence of their hero. "If I could catch the man or woman who is giving you all this botheration," wrote R. M. Barrington from far-off County Wicklow, "I would unhesitatingly give him or her a cold shower bath – in their best clothes – and douche them again and again until they apologized to you." "*You* are the 'Father' of the Club, which is and has been the only connecting link between the Botanists of Great Britain", ran another fanatic's outburst. As for Moss, a certain W. A. Harford, of Bristol, had a topical solution to suggest: "Send him out to the Dardanelles ... he must be a *German*." Chauvinism came cheap in that summer of 1915.

Greatly reassured by such boisterous messages of support, Druce set about composing the necessary rejoinder to Moss. Realising that his accuser had done himself no good at all by the polemical line he had taken, he was careful to adopt a subdued and dignified tone, contenting himself with recounting the history of the Club, in the belief that his policies would speak for themselves when viewed in the necessary perspective, and with reporting the verdict of the replies to his questionnaire, which told so crushingly in his favour. The resulting document went out somewhere around the middle of the autumn.[29] It appears to have occasioned very little comment; for by that time it was clear to all concerned that Druce's opponents had lost the struggle: his support among the newer members was simply too strong to make it possible for him to be reined in or dislodged. All that remained to be done now was for an armistice to be declared and for the weapons to be removed from the battlefield.

On December 4th Moss issued a brief, final circular formally conceding defeat. Only twenty-eight members, he reported, had replied in favour of his proposal, while about half, despite having been sent stamped and addressed postcards, had returned no answer at all. "Not all the active malcontents were in favour of my suggestion which, I gather, was too moderate for some of them," he could not refrain from adding. In view of this result he now had no option but to keep to his declared intention and resign from the Club. "If and whenever it becomes a democratic organisation," he concluded with a Parthian shot, "I shall, of course, be pleased to consider the question of rejoining it."

But that question was never, after all, to arise for him. Inside a year, despairing of his prospects at Cambridge, with his income almost halved by the wartime shortage of students and with his marriage breaking up in the scandal of a divorce, he decided to make a new start abroad. A chair in South Africa happening to fall vacant, he applied for it and was appointed. Even as the ship that was to take him there set sail he was still continuing to splutter at Druce's "fiendish vindictiveness", "cringing sycophantism" and "diabolical untruthfulness".[30] The *Cambridge British Flora* was left under his nominal editorship, but the main burden of carrying on that work was delegated to a young ex-student of his who had recently joined the staff of the British Museum (Natural History), Alfred James Wilmott. Wilmott performed very creditably, but the *Flora* failed to pay its way and in 1920, much to Moss's intense chagrin, the publishers decided it would have to be discontinued. It was the final banging shut of the door. The one man who had promised to bring to British field botany permanently more rigorous standards and a larger scientific purpose, the one man at that period capable of welding the amateur and professional wings in productive new enterprises, had withdrawn his services utterly and conclusively. British botany was never to set eyes on him again.

G. Claridge Druce, Secretary–Treasurer of the Botanical Society and Exchange Club.
Oil painting by P. A. de Laszlo, subscribed for by the members in 1930.

CHAPTER 9

King Druce

That meeting 'after the War' at which everything was to be regularised never in the end took place. Indeed, nothing was to be heard of the proposal again. Fearful of confronting his critics, some of whom were his long-time friends, Druce seems to have quietly buried the suggestion; while the enemy contingent, for its part, no doubt decided it was pointless to pursue the matter: Druce would only pack such an assembly with his supporters and no one would be prepared to brave their hostility and denounce his behaviour to his face.[1]

This was all very understandable; yet the inaction was a serious mistake. For until and unless there were meetings, there was no possibility of restoring even a semblance of democracy. And without that many of the foremost workers in British systematic botany, whose support the BEC needed if it was ever to become a national society of any scientific weight, would either abstain from joining or else continue to belong to it in no more than a nominal capacity. H. W. Pugsley, for example, by general consent probably the best of that generation of amateur taxonomists, was already holding aloof and would go on doing so throughout the Drucean ascendancy, giving much help to the Watson Club instead and publishing his work in the *Journal of Botany* or in one or other of the Linnean Society periodicals.

Moreover, contrary to what many later came to suppose, Druce had started out with every intention of convening such a meeting: it was not an airy assertion he had made simply to appease the opposition. Plans for such an occasion had in fact proceeded quite some way during the winter preceding the War (on the evidence of letters to him from members, some of whom were enthusiastic). What he had in mind was something along the lines of the present-day Recorders Conferences, with talks interspersed with business and field excursions. These last, he suggested, might be in conjunction with the Society for the Preservation of National Areas[2] and the Ashmolean NHS of Oxfordshire[3] – for as the venue he naturally proposed Oxford, which he could claim convincingly enough as a convenient centre. A notice advertising the event for that July was circulated during the spring and Bailey was even approached and invited to preside;[4] however, "so few members sent in their names as wanting to attend that reluctantly it had to be deferred", Druce later reported.[5] It was rescheduled for June 1915 and a chairman sought for that time in turn, though equally vainly, in the person of W. P. Hiern, the leading Devon botanist and authority on Water Crowfoots.[6] Thus a sufficiently solid basis had been established on which a more determined set of members could have insisted on building. Druce himself, too, might well have succumbed to the pride of playing host, had he only been pressed on this.

It would clearly have helped had there been an established tradition in British botany of meeting on a national basis. But unfortunately, apart from the Edinburgh and Linnean Societies, which had come to be the focus essentially of the residents of the respective capitals, the sole continuing example of this was provided by the British Association for the Advancement of Science, and the very broadness of that body doubtless obscured its appropriateness as a model for botany specifically. All other efforts had long been left to the local societies, either on their own or in concert with their neighbours in the various regional unions which had gradually been emerging. Even so the meetings of the latter constituted a fair approximation to what the BEC required, and as Druce had had first-hand experience of organising those of the Midland Union of Scientific Societies he could scarcely plead unfamiliarity with the pattern.

In the absence of such a tradition, the leading amateurs were hard to draw away from home. They were just not in the habit of making long journeys for the main purpose of meeting one another: such travelling as they were prepared to undertake – or could afford – was reserved for exploring new parts of the country and thereby expanding their collections or at least their field experience. It was not the inadequacy of transport that was in any way to blame. This, in any case, had lately improved enormously, as motor vehicles had arrived to supplement the long-existing trains. Druce had acquired a car by as early as 1909 and even a poorish country vicar like Marshall had followed his example by 1914: "it comes in handy for botanical outings, beyond carriage-drive range", he reported enthusiastically.[7] By the outbreak of the War the first local society had even made the switch to motor coaches for its excursions.

To a limited extent this transport revolution eventually forced Druce to a compromise: for in 1923 that suggestion he had made all of fifteen years earlier was at last taken up and an annual field meeting instituted. Admittedly, he did not have to do the organising – that was undertaken by Miss Eleanor Vachell – and it was left very vague whether this activity was officially a BEC one or not; but he did turn out each time to act as Leader. Once again, therefore, his repeated excuses that the members would not attend a meeting if one was called must be regarded as hollow.

That first-ever 'excursion' (the preferred term for a field meeting in those days) was to Jersey, a part of the British Isles that could be guaranteed to produce an ample turn-out. It lasted five days and most of the island's well-known rarities were visited, "one feels certain without damage being done to any ...., as most of the members were contented with painting them", wrote Miss Vachell in a lengthy account published in that year's Report.[8] Predictably, the party hired its own transport: "a heavy shower ...", we read, "only damped the clothes and not the ardour of the flower-hunters, who cheerfully lunched under the partial cover of the charabanc".

The success of this experiment led to further excursions in subsequent years. On the 1924 excursion the party saw, or liked to think it saw, "all the rare plants that occur in Caernarvonshire and Anglesey", again travelling by coach and hiring a special train for the ascent of Snowdon. The report on that occasion reveals that Sunday mornings were kept free of any fixture, presumably to allow those who wished to attend morning service.

It is not clear whether these primordial field meetings were advertised among the membership generally or whether they were by invitation only – one suspects the latter. The report on the 1927 Excursion is in this connection doubly revelatory, referring to it, delicately, as a meeting "of some of the Botanical Society of the British Isles". The omission of the words 'and Exchange Club' was almost certainly pointed (rather than mere shorthand) and hints at the reason for the exclusiveness: it was to keep away the collectors. From other evidence it would appear that most of those who attended these occasions were women, and women moreover who subscribed with some zeal to the code of non-picking associated more especially with the Wild Flower Society, that 'botanical nursery' (as Druce is said to have termed it) founded back in 1886 from which the BEC had lately been recruiting more and more heavily. The *raison d'être* of that Society was, and still is, the competitive sighting and logging of the maximum number of species within the British Isles, typically on a yearly basis, and there was a powerful demand from those of its members who belonged to the BEC for conducted tours of the areas richest in rarities under experts able to name what was seen. Indeed, as Druce probably realised, unless the BEC was prepared to offer such a service, it would probably not have been able to attract or retain much of this sizeable extra contingent.

That it was Druce who led these meetings, however, was heavily ironic. For, as most people were aware, he was one of the most voracious of collectors. Keen conservationists quickly learned that he was not someone who could be safely shown their treasures. W. H. Mills was once ill-advised enough to take him to see the Cambridgeshire dactylorchids and was horrified when Druce pulled specimens up by the armful; he would have hit him, he recalled, had not their companion, A. H. Evans, forcibly restrained him.[9] But in fairness it must be appreciated that Druce belonged to an older generation that was inclined to be heedless in this matter. Marshall, for instance, punctilious enough when asking to be directed to the Monkey Orchid, *Orchis simia*, to promise "not to take roots or in any way injure the station",[10] nevertheless saw fit to help himself to sufficient of the dangerously rare Blue Heath, *Phyllodoce caerulea*, to cover at least six sheets in his herbarium.[11] Under pressure from the many new members who held strong views on this subject, Druce was placed in an awkward position. In his Secretary's Report for 1915 he felt obliged to strike a stance of progressiveness: "It is a matter of satisfaction," he purred, "to know that all our members are impressed with the importance of avoiding reckless or careless gathering of plants which endanger any local species, and that they are anxious to do all that is possible to protect them from injury." But at the same time, all unwittingly, he exposed this for the humbug that it was by including in that very same issue a request by F. J. Hanbury for the "seeds or roots of rare British species". Despite this blind spot on collecting, though, he did a great deal of work in these years for the newly-founded Society for the Promotion of Nature Reserves, providing that body with all of its botanical guidance on the sites most worthy of protection. Evidently there was a clear-cut dissociation in his mind between the removal of specimens and the safeguarding of their habitats. It may be that, like G. A. O. St Brody, the Gloucestershire botanist, he lived under the fond delusion that the power which causes a plant to occur in any given spot in the first place can be depended upon to preserve it there in perpetuity.[12]

Yet it was perhaps as well that Druce was personally muddled on this issue, for he and only he could have held together in the one organisation two such mutually hostile sets of people. "I am really glad to hear there are so few Exchange members", wrote in one of the newer members, aggressively, in 1915;[13] yet this was to the man who only four years earlier had effected a sudden near-doubling in the size of the Distribution, to the extent that after a long period of being half as large again as that of the Watson Club it temporarily became up to four times as great. In all probability this increase was merely one more result of Druce's general expansionary energies: it is unlikely that he had in mind anything so deliberate as a compensatory offering designed to keep within the Club the more scientific minority.

It is tempting to see these two opposed groups, the collectors and the anti-collectors, as corresponding very broadly to Druce's opponents and sympathisers respectively. Yet that is over-simple. While it is certainly true that all field botanists in that generation who aimed to do serious scientific work found the forming of a herbarium of reference more or less obligatory, by no means all of these were on uneasy terms with Druce. A. R. Horwood, for example, the leading Leicester botanist and a most industrious collector and exchanger, was devoted to him. So was W. H. Pearsall, the foremost specialist in aquatics. So, too, was so unlikely a person as W. B. Turrill, the leading British pioneer of experimental taxonomy and later Keeper of the Kew Herbarium: Druce had given him an early helping hand and was able to persuade him to act in 1928 as the Distributor.

Druce's support came in fact from a remarkably diverse constituency. As a self-made man and a dedicated Liberal, his *rapport* with the less well-placed was only to be expected. What was much less expected was the size of his following at the loftiest levels of society. Some of this he presumably owed to his prominence in local politics: one partly-botanical gathering described in his Secretary's Report for 1928 is noticeable for having included an Asquith, a Churchill, a Stanley and a couple of Liberal Privy Councillors. Some, too, he must have acquired through his long and intimate associations with Oxford University (of which foreign correspondents often assumed he must be one of the professors). Most of it, though, probably came about through the simple accident that that world of the week-end house parties needed its tame flower-naming expert. 'Painting one's Bentham' is known to have been a fashion in débutante circles in the Twenties[14] and the evidence seems to suggest that this fashion had been inherited from those débutantes' aunts and mothers. Druce was endlessly patient with beginners; he was very well-read and, in later years, widely travelled; he was renowned for his fund of drawing-room stories; years of cultivating custom as a tradesman, moreover, had given him delicate antennae and an ability to charm at will. One entrée had led to another and gradually he had been absorbed and accepted.

It was natural therefore that he should have treated this as prime BEC recruiting territory. Very helpfully, it was made up of great chains of friends, who were only too willing to propose one another as members and who could afford to be heedless of 'just another small subscription'.[15] Druce, moreover, can hardly have been unmindful of the advantages in those days to any social grouping outside the strict domain of learning of a show of prestigious names. One of the great triumphs of his youth, indeed, had been the launching of the Northamptonshire NHS in the teeth of fierce social-cum-sectarian animosities by his securing as founder members influential representatives of each of the various rival groups.[16] More simply still, though, as Rendle wrote in one of his obituaries of him, Druce "dearly loved a title".[17] Among his botanical friends he was renowned for this foible and among his surviving letters there are several from them as far back as 1909 chaffing him on this score ever so gently.

The influx of 'outsiders' of which Marshall and others had complained so bitterly had a most unusual tinge to it in consequence. To an extent which must surely be unique in any learned society in modern times, and certainly in any scientific one, the *hoi polloi* which proceeded to pour in consisted – for a change – not of plebeians but of the aristocracy. Suddenly in 1914, titled members, who had been almost non-existent a year or two before, constituted one in every ten. And their numbers continued to rise steadily until the end of the 1920s, by which time they accounted for one in every eight.[18] This gave the BEC the appearance of an eighteenth-century throw-back, with cars merely exchanged for coaches, charabancs for barouches. To add to this impression a number of wealthy *grandes dames* appeared on the scene, most notably the awesome Mrs Wedgwood, and bestowed their bountiful patronage on a succession of young botanists of promise.

To an even more pronounced extent the influx had also brought a change of gender. All through the years the Club, to its credit, had always had one or two women among its members; but as soon as Druce took over their proportion began to increase noticeably. By 1914, when the membership as a whole had already enjoyed six years of very rapid growth, women numbered one in every six. By the end of the 1920s the ratio had narrowed to nearly one in three and it was to continue to narrow further all through the 1930s. Such an inexorable rise would be impressive even if the size of the BEC had stayed static: as it was, from 1908 to 1928 the published membership figures show that it soared sixfold and then dwindled sharply.

Unfortunately, though, as was to come to light only many years later, the membership figures at this period are not at all to be relied upon. Druce fell into the same trap as his BSL predecessors: in his anxiousness to see the total mounting higher and higher he refrained from deleting from the lists those who failed to keep up their subscriptions. Many, it later transpired, never paid for years, just as in Bailey's time. To add a further element of falsity, some who had served the Club well in past years (such as the Revd W. Moyle Rogers, long the referee on Brambles) were kept on the books in an honorary capacity. Druce also believed in meeting out of his own pocket, or out of the specially-created Benevolent Fund, the subscriptions of promising young newcomers or the especially impecunious. The absence of a separate Treasurer made it easy to float the Club, much as Bailey had done, on a raft of well-concealed benevolence.

It was surely this freedom of operation that he enjoyed which, more than anything else, led Druce to prolong indefinitely the disgraceful constitutional vacuum. He was able to run the Club's affairs immune from any interference – no small consideration for someone as touchy as he – and with the minimum of formality. The subscriptions did not even go into a BEC account at the bank: Druce found it simpler to pay them into his personal one instead.[19] In the same way the Club's possessions were all intermixed with his own, a practice which was to cause much difficulty when the time came to separate them after his death.

The one person on whom he had no choice but to rely, fortunately for him, shared this preference for hidden subsidies. This was R. H. Corstorphine, the printer to the Club. A former chemistry teacher and a keen field botanist in his own right from student days, Corstorphine had married the daughter of the proprietor of T. Buncle & Co., a small firm of printers-cum-publishers in Arbroath, near Dundee. On the inheritance of this by his wife in 1899, he had become its managing director and turned it into the leading specialist in Britain in publications involving the scientific names of animals and plants. He ran it less as a business than as a hobby, it is said, and certainly its prices were extremely competitive. Deeply loyal to Druce and to the BEC, he deliberately charged the Club less than the proper commercial rate over a period of many years, strictly forbidding this ever to be revealed.[20]

Corstorphine's indulgence, combined with various donations which Druce was able to raise from time to time from friends or wealthier members, explains how it proved possible for the Reports to swell so greatly in size in the post-war years without any corresponding increase in subscription. That for 1928, in particular, ran to an astonishing 339 pages with a supplement on the Flora of West Ross which ran to 149 more. The entire bill for this was shown in the accounts as covered by 'grants' to the Publication Fund of £220 which had materialised conveniently from some unexplained source, mainly, one suspects, Druce's personal exchequer.

The expansion in the Reports indeed was such that even Druce was unable to supply all of the material himself any longer, even supposing he still wanted to: his wish now, rather, was to see them develop into that journal given over exclusively to British systematic botany which Watson and Syme had planned unsuccessfully all those years ago. Alas, though, the role of editor was not at all his forte: he was too anxious to encourage, too reluctant to disappoint: everything sent in, it would seem, was accepted and printed, not even excluding verse. A good deal of what appeared was barely scientific at best and written in a style more suited to a parish magazine. This deliberate lightening was clearly popular with the general run of members, but a heavy price was paid for it in the alienation of many of the more scientific of the country's amateurs and of virtually every one of the professionals.

Yet it was not a good period for taxonomy, in Britain or anywhere. The classical approach was almost played out, unable to clarify any further all but one or two of the traditionally complex genera and groups; but the new methods, the crucial extra insights that were to be provided by the geneticists and the cytologists and the biometricians, were as yet hardly born and in any case deployed in a secluded Academia. To read again today the literature of field botany of those years is a profoundly depressing experience. So many of the monographic revisions were basically flawed, so much effort went into piling up records of entities later exposed as without substance. All those 'splits' by C. E. Britton in *Centaurea* and *Melampyrum* . . . all those extra species in *Viola* laboriously introduced by Eric Drabble . . . the vast, utterly artificial edifice erected by A. H. Wolley-Dod in *Rosa* . . . The specialists of the day did their best, but in too many cases the definitive tidying-up of these long-obscure corners was still out of reach in default of an understanding of the mechanisms that lay behind the variation.

It was not only British workers who were being self-deceived. In Austria, in just the same way, *Thymus* was grossly overloaded with species by Ronniger, as in Sweden *Capsella* was by Almquist. Yet Continental work was still regarded in Britain with an unhealthy degree of awe and the pronouncements of specialists such as these tended to pass without challenge. For this, Druce's own uncritical habits were also much to blame. The attitude he persisted in maintaining was that of the old-style herbarium botanist, with his keenness to 'get a name' almost without regard to its soundness. He could never resist a supposed new authority, delighting in being first to shoot off his material and first to receive back a parcelful of gratifyingly unfamiliar determinations. Though he collected in the critical groups conscientiously and copiously, he never had the inclination to study any of them in detail; he was therefore necessarily indiscriminate in choosing between rival treatments. His bias in such cases was towards the radical upstart, the overturner of the established – but this was an emotional bias, not grounded in scientific judgement, and sometimes it could lead to trouble. When the youthful W. C. R. Watson appeared on the scene, for example, he freely sent him all his folders of brambles and published his verdicts in the Report; but this was all done without reference to Riddelsdell, the then acknowledged national expert, who up till then had never even heard of Watson, found his determinations extraordinary and was understandably cross.[21] It does not seem to have occurred to Druce that putting into currency conflicting interpretations of a group was bound to have a retrograde effect by undermining confidence in both.

Another victim of Druce's uncontested monopoly was the standard in the publishing of vice-county records. At the time he succeeded to the Secretaryship the system inherited from Watson in this respect was being adhered to faithfully, and Arthur Bennett brought out, in the *Journal of Botany* in 1905, the first of what was generally assumed would be a whole series of supplements which periodically updated *Topographical Botany*. In his overindulgent way, though, Druce began to admit to the Club's Reports records that had merely passed his own, often rather cursory scrutiny – records which were not necessarily erroneous but not necessarily new either and which were published, it seemed, merely to please those who had submitted them. In 1917 he had canvassed the excellent idea of reviving the system of Local Secretaries maintained by the old Botanical Society of London and giving these the express task of keeping the records for their respective vice-counties up to date; but although several correspondents[22] reacted to this suggestion with enthusiasm, it shared the fate of everything else which would have entailed some delegation from the centre. Instead, Druce insisted on taking upon himself alone the formidable task of compiling a successor to Watson's great compendium and, predictably, he put off repeatedly the final writing-up of this till the very last months of his life, when his powers had begun to fail. *The Comital Flora of the British Isles* was eventually published, at the BEC's expense and at a cost that was really beyond it, on the day following his death – and to a chorus of denigratory comment.[23] "An entirely useless book", "so incorrect that the copies ought to be called in and burnt": these were reactions typical of many. Druce had foolishly departed from Watson's crucial plan of indicating, by appropriate abbreviations, the grounds on which each vice-county entry was included. His excuse was that this would have been too costly and the space required enormous. Contrary to general belief, though, he did preserve the fuller manuscript from which the published work was abstracted; many years later, in 1942, a copy of *Topographical Botany* cut up, pasted into a ledger and annotated in Druce's handwriting was found among the BEC's property at his former home in Oxford.

All the same Druce's personal fondness for this now-traditional form of botanical bookkeeping did ensure that many interesting finds were captured in print which might otherwise have gone unrecorded. For the web of correspondence which he had spun over the years and of which he now sat at the centre was a very extensive one indeed, and his willingness to name specimens for anyone, however inexperienced, was by this time legendary. "If the plant is too abstruse, pack it off to Doctor Druce" had acquired the status almost of an incantation. And needing little sleep, Druce would toil away far into the night, patiently examining the contents of envelope after envelope and dashing off his conclusions – and maybe his congratulations – in that famous, near-indecipherable scrawl.

This is the Druce who has remained most firmly in the memory of those who lived on to tell of him afterwards: the Great Expert who was never too proud or too busy to help even the most foolish with a query. Aptly venerable in appearance, he had gradually been promoted into a living symbol, a magus figure, the recipient of a mixture of awe and admiration of cult-like intensity and proportions. For some, even, who had known of him only by repute an eventual encounter in the flesh had something of the impact of a visionary experience: "Had it been the Apostle Paul I could not have been more ... delighted," one of them was later to record: "the name Druce added several cubits to his stature ... I felt half-afraid of him, this man who was born on the same day of the year as Linnaeus."[24]

As Druce's eightieth birthday approached, in the spring of 1930, the glorification rose to a crescendo. Laurels, too, now arrived for him of a greater and greater splendour. An Oxford DSc, contrary to general belief obtained by him by examination, was joined by an Honorary LLD from St Andrews, that same university which had similarly honoured Syme, back in 1875, on the completion of *English Botany*. And in 1927, very much more remarkably, he ascended to that pinnacle of scientific eminence, election as a Fellow of the Royal Society – an astonishing feat for a largely self-educated amateur in what had become by then very much a world of the professionals.

All too little of this grandeur rubbed off on to the BEC, though Druce did his best. Undoubtedly he would have dearly loved to have been able to rename it the *Royal* Botanical Society; but that title, unfortunately, still continued to be pre-empted and he had to content himself with securing H.R.H. The Princess Royal as Patroness instead.[25] In similar vein, but with a touch of bathos, he concocted for the Club a crest consisting of the Giant Water-lily of the Amazon – chosen for its particular association with the ancestral Botanical Society of London – with its (actually, illegitimate) name VICTORIA REGIA – chosen for its royal ring – beneath it. Complementing this was FLOREAT FLORA as the uninspired motto. Despite the absurdity of a body concerned only with the British flora having as its emblem a tropical species, the use of this was to persist, albeit with less and less conviction, down till very recent years.

For all this gilded surface, though, the BEC remained perilously insubstantial underneath. It was very largely one man's creation; and when that man was gone, it was not at all certain that it would manage to survive. By turning it so fully into a reflection of himself, Druce had put at risk inadvertently the object of all his labours.

On the other hand if there had been no Druce, would the Club have lasted as long as it did? But for his dash for growth it would surely have kept on its way as a deliberately tiny clique and eventually petered out when the climate turned against collecting. For all his failings Druce had at least bequeathed to British systematic botany the necessarily much-enlarged framework around which a broader and weightier society could potentially be erected. That this framework had been jerry-built, that it was a good deal less solid and more fragile than it appeared to the outside observer, was really beside the point: what mattered was that it looked imposing. It seemed too sizeable now to be bypassed and ignored. Instead of building anew, therefore, the many able botanists who had held themselves aloof from the BEC so long as it was 'Druce's society', decided to transform the existing shell into the worthier kind of structure which they had long sensed that their area of the science needed. But the conversion was to prove a less simple task than probably most of them had imagined, and what in due course emerged retained among its features more than they can have intended that was obstinately Drucean. The influence of 'the Doctor' was not to be easily shed.

CHAPTER 10

Return to Democracy

On his death, in February 1932, Druce was found to have left the enormous sum of £90,000 – at present-day values about two million. Not one penny of this, however, was bequeathed to the Society (as the BEC now tended to call itself), at any rate explicitly, despite the central place it had occupied in his life through all his later years.

It was clearly not a case of an old man's absent-mindedness, for there was a long list of minor legacies to relations and friends as well as to a variety of charitable causes. Even the BEC's counterpart, the Botanical Society of Edinburgh, received a token £100; and a more than comfortable £600 was earmarked for the production in his memory of a volume of life and letters.[1] Otherwise, it turned out to be his wish that virtually everything was to be divided between Oxford University and the Society for the Promotion of Nature Reserves.

However, the BEC was not overlooked in the will entirely. In leaving his house to the University along with his herbarium and library and a special £12,000 endowment, he expressed the desire that this should be called "The Claridge Druce Herbarium and Institute" and that his young librarian and secretary, John Chapple, who had been in his employ since leaving school, should be appointed its first Curator. Chapple, in turn, he wished "to carry on as far as possible work at British Field Botany" and "undertake to give adequate assistance to the Botanical Society and Exchange Club, if so desired by that body". Thus as a by-product of assuring his collections of a permanent, cared-for and (as he hoped) continuously useful existence by vesting them in an institution devoted to teaching and research, Druce ingeniously aimed to bequeath to the Club the not inconsiderable gift of a salaried secretariat.

Very welcome though this was, it must nonetheless have come as a grievous disappointment. Even a small fraction of Druce's money would have been enough to transform the Society's fortunes utterly and lift its eyes in altogether more ambitious directions. He himself must surely have seen this, and one can only guess at why he stopped short of helping it more directly and more munificently. As the Society as a physical entity was more or less coterminous with his house and a proportion of its contents, it may be that he felt that the provision he had made offered the only sure means of assisting it, in the absence of any guarantee that it would continue to flourish and be capable of putting to good use a large financial windfall. At the back of his mind there may even have lurked the fear that the Society might pass in due course into hands that were unsympathetic and that any sum he bequeathed to it might then be used in some way or other inimical to his memory. By tethering everything to Oxford he could at least be sure of preserving the close identification between himself, the BEC and his adoptive city.

The University, however, soon made it clear that it had no use whatever for the institute he had envisaged. Interest in systematic botany there had died out more or less completely and it was hard to see any call being made on its facilities by either the taught or the teachers. Hopes accordingly began to grow that it might turn down the bequest and that the institute, after being offered in turn, equally vainly, to the British Museum (Natural History) and Kew, would end up by coming to the Society after all. Alternatively, it was speculated, Druce's executor, Westminster Bank, might see fit to retain it but entrust its running to the Society indefinitely.[2]

The question was, though, whether the £12,000 endowment went with the institute or was bequeathed to the University regardless. At the depth of the Depression, when funds were exceptionally hard to come by, this was by no means a trivial sum for even Oxford to forgo. If the only way of obtaining it was to give the white elephant house-room, then it was prepared to stoop to this; but its strong preference was for the money on its own unencumbered.

Unfortunately, Druce had modified his intentions so frequently that his solicitors had failed to keep abreast of them.[3] There was a whole series of testamentary documents, it turned out, any one of which might be the definitive version that qualified for probate; the wording of these, moreover, was dangerously loose and open to varying interpretations. The only way such uncertainties could be resolved was by seeking a ruling from the courts. But, as so often, what looked at first like a comparatively simple matter, requiring perhaps one brief hearing at most, was found on closer scrutiny to be a tangle of complexities and the case dragged on and on for what proved in the end to be several years.

As a potential major beneficiary the Society thus found itself left hovering, in a state of chronic expectancy, unsure as to which of the very different outcomes to plan for. At the same time, more immediately, it was placed under enormous strain administratively, for until probate could be obtained, which did not come till January 1934, the entire estate remained frozen.

Druce's casualness about what was his own property and what was the Society's resulted in the executor necessarily taking over everything and releasing items only where it could be demonstrated that ownership resided elsewhere. Because the subscriptions had gone into Druce's personal account, these continued to pass into the custody of the executor, who then had to be asked to forward them to the Treasurer, an absurdly elaborate procedure which persisted till as late as 1937 or even 1939.[4] The executor, similarly, had to be asked to approve every request for loans of material from the herbarium. The chief sufferer, though, was poor Chapple: until the estate was unfrozen, there were insufficient funds available to pay him more than a minuscule £100 a year.[5] Early in 1933, finding him "in rather desperate straits", the officers of the Society took pity and advanced him £50 out of its own proportionately slender purse. The following year, when his position was regularised at last, it was decided that it would be only right and proper for the Society to pay him a standing honorarium. Accordingly from that point on he received from it £10 yearly.

Even after the will had been proved, it was several years more before it was finally settled that the University would have to accept the institute if it was to receive the money. As a result it was 1939 by the time the two residuary legatees eventually came into their shares.[6] The War then intervened; so that even as late as 1945 the BEC had still not established with the University whether it was going to be allowed to continue to keep its possessions in what it had come to regard by then as its semi-official office. Its barely visible presence was in the event tolerated for some years further, and it was only in the mid-1950s that the link with the University, and so with Oxford too, was finally and conclusively severed.

Fortunately, all through these years of legal limbo the Society had plenty to concern itself with internally.

The first and most obvious priority had been the speedy dismantling of the Drucean monarchy. Druce himself had airily assumed that the kind of regime he had established would continue after his death, the splitting up of the Secretary's functions being the sole modification that was needed. To this end he nominated a so-called Advisory Committee, consisting of twelve leading members, with the express task of choosing his successor. The names of these were sprung on them in the 1929 Report without some at least of them having even been consulted.[7] They comprised in the main staunch supporters of his, such as W. H. Pearsall, the schoolmaster editor of C. E. Salmon's posthumously published *Flora of Surrey*, and R. H. Corstorphine, the natural history publisher and printer mentioned previously. Also included was one little-known figure who had evidently been selected as an *éminence grise* rather than on account of his standing as a botanist. This was the much-respected H. T. Baker, Bursar and later Warden of Winchester College, a Privy Counsellor from a pre-war spell as Liberal MP for Accrington, and a one-time President of the Oxford Union (in addition to which, unbeknown to anyone then, he had spent the War in the loftier layers of secret intelligence, as assistant to the novelist John Buchan in the newly-named MI5).[8] In the next year's Report Druce took a step further and went so far as to designate Pearsall to succeed him as Secretary, Chapple to become Assistant Secretary, a capacity in which he had already been serving for some time unofficially, and T. J. Foggitt (son of the Foggitt who had sustained the Thirsk Club at the time of the disastrous fire) to take on the prospective new office of Treasurer jointly with his wife who, as the former Miss Gertrude Bacon, had long been a prominent member in her own right. Pearsall later revealed that he had already promised Druce some years previously that, should circumstances make it necessary, he would step in and take his place.[9] His choice as heir-apparent he himself believed he owed mainly to the fact that Druce considered him 'safe' on the matter of not allowing anything objectionable to appear in the Reports;[10] but it cannot have been irrelevant that he was also one of the most fervent of Druce's admirers and had reviewed two of his books in terms that were shamelessly uncritical. Now for the second time, in the last year of his life, Druce approached him and secured his agreement to taking over when he was gone, at any rate as a temporary measure.[11] At the same time, in common with the rest of that shadow cabinet, his views were solicited on how the BEC could best be run in the future.[12] No one can say that Druce showed heedlessness of what would become of his creation.

In the event things did not turn out altogether as Druce had planned. Within just over a fortnight of his death the Advisory Committee was indeed convened and it duly fulfilled its allotted role of inviting Pearsall to assume the Secretaryship. But it then took a sharp turn and decided that the BEC should revert at once to democratic procedures. What was said on that occasion to induce the twelve apostles to make such a radical change of front we are unfortunately unlikely ever to know, for no record of the discussion appears to have survived. It seems highly likely, though, that the foremost advocate of the change was Pearsall himself. In a letter he sent just a week or so before to one of his fellow Committee members he had urged that "our first aim should be to constitute a *Society* on the usual grounds", by which he meant one with an elected set of officers. Only in this way, he contended, could they remove the stigma of the BEC's questionable existence as a legal entity in the eyes of the bodies, in particular Oxford University, with which it was, or was likely to be, necessary to negotiate.[13] By a supreme irony, it seems, Druce's testamentary muddling had had the last effect that he may have intended: the immediate inversion of his system of governance.

In accordance with the Advisory Committee's decision, a General Meeting was thereupon called to allow for the initial elections to office. This took place some two months later, on May 9th 1932, in the Linnean Society's rooms in Burlington House, henceforward the nearest the BEC, and the BSBI after it, were to have to a headquarters. The fact that it was arranged for a Monday afternoon may have reflected a wish to discourage a large attendance; more probably, though, such a time was no more inconvenient than any other for that high proportion of the members who were untrammelled by a profession.

Inauspiciously, the last general meeting that had been held had been to wind up the Botanical Society of London three-quarters of a century earlier. This one, however, seems to have passed off in an atmosphere of optimism and goodwill, albeit with a fair degree of tension just beneath the surface.

The reason for this tension was the coming together at last of Druce's supporters and opponents. To achieve a lasting reconciliation between the various warring factions into which British field botany had become so damagingly split was an obvious priority for the newly-reconstituted body. To this end the names put forward for election to the General Committee, as the equivalent of today's Council was initially termed, were carefully selected in order to ensure that the main rival interests would all be represented. Fortunately there were ample places to allow this, for its size had been set at a more than generous eighteen (exclusive of the officers).[14] Alongside the familiar names of Druceans there consequently now appeared a number of quite fresh ones: R. W. Butcher, W. A. Sledge, A. E. Wade, J. Ramsbottom, A. J. Wilmott and J. S. L. Gilmour. Through these new recruits Kew, the British Museum (Natural History), the university world and the provincial museums all received some coverage; indeed, approaching half of the new Committee consisted of professionals. In this one fact alone the new body demonstrated that this was to be a very different BEC from the one that had preceded it.

This heavy weight of professionals was all the more necessary for the reason that it had been found tactful to vote into the respective offices all four of Druce's nominees for them. Of those, Pearsall and Mrs. Foggitt in particular, were strong characters and some equally forceful individuals of an offsetting persuasion were obviously called for with a view to keeping them in check. Wilmott and Ramsbottom, the two British Museum representatives, fulfilled this requirement perfectly: both were former protégés of Druce's arch-enemy, Moss, and had conspicuously disassociated themselves from the Club and all its works throughout the preceding years. In Wilmott's case there was the further countervailing pull of a special, lifelong friendship with R. H. Goode, the long-time Secretary of the increasingly antagonistic Watson Club. Wilmott, moreover, unlike Ramsbottom, who was a mycologist and so belonged to a different botanical sub-community, had the taxonomy of British vascular plants as his exclusive professional commitment; he could therefore be counted on to uphold the strictest standards over what was accepted for publication in the Reports.

With so much potential friction built in from the very start it is at first sight surprising that the new regime functioned as smoothly as it did. For this, it had the inspired choice of H. T. Baker as Chairman to thank.[15] As one who served under him was later to record, "his standing commanded respect, and his quiet dignity, great patience, and good humour earned it . . . The fact that he made no attempt to do scientific work freed him from any temptation to take sides in the controversies of the day. His sympathy and charm, combined with quick perception of the fundamental problems involved, often solved differences of opinion almost as soon as they arose."[16] In the end Baker was to serve in this key moderating role for as long as fifteen years, seeing the Society through to the end of the Second World War.

It appears from contemporary letters (for the minutes have not survived and this was not the kind of detail to find its way into the Reports) that at that time the General Committee met three times a year. Patently, that was not sufficiently frequent for the deciding of urgent matters; the very largeness of the Committee, moreover, ordained that its discussions would be ponderous and perhaps inconclusive. It was accordingly agreed that, as in most other societies, there should be an official inner core, to be named the Executive Committee (the equivalent of today's Co-ordinating Committee), which could be called together at short notice and would have the power to act on anything pressing. For obvious reasons, Pearsall urged, the members of this ought all to reside near London;[17] for the metropolis had now, once again, become the Society's unquestionable centre of gravity, as it has been ever since.

William Harrison Pearsall, who had now emerged as the Society's main driving-force, was an ideal 'bridge' figure. As Druce's anointed, he automatically commanded the allegiance of the main mass of the membership; at the same time his reputation as a field botanist, while inflated (for his

James Britten.

C. E. Moss.

W. H. Pearsall.

work on the Pondweeds *(Potamogeton)* all had to be done again, after his death, by Dandy and Taylor), was sufficient to make him acceptable to the amateur élite and the professionals. Like Druce he was retired and so free to devote most of his time and still-abundant energy to the Society's affairs, for though he was over seventy by then he gave the impression of being far younger; like Druce, too, he had had extensive experience in municipal politics, including the chairmanship of his local council finance committee. Much of his life, though, had been passed in too small a pool and to the conceits he owed to that he had unfortunately added the particular professional vices to which headmasters are prone: a strong streak of vanity and a certain sanctimoniousness. He quickly mishandled the most difficult of his fellow officers, Mrs Foggitt, while he was inclined to order about the youthful Chapple, instinctively, like one of his former prefects. This did not make for a contented and united team. Nor did it help that, in his belief in the all-importance of the personal touch that Druce had brought to the Secretaryship, he strove to project himself in a similarly obtrusive fashion. Had he only confined this to his answering of members' botanical queries (a much more onerous inheritance, he was heard to confess, than he had imagined), no one would have objected; but the undue parading of his name on the cover of the Reports and his over-pervasive editorial presence proved an irritant to many. And he annoyed even more by high-handedness in the publishing of plant records, passing over inexplicably many of those submitted by the experienced while keeping up Druce's bad practice of including many others of scant interest merely in order to give heart to novices.[18] At least one major contributor took such offence at this that he switched to publishing elsewhere, at any rate for a time.[19]

Behind Pearsall's policies there lay the haunting worry that the BEC was on the edge of an abyss. A high proportion of the members, he was only too aware, had joined simply out of personal affection for Druce and there were ominous signs that these would all now melt away. Nearly 100 had indeed done so immediately following Druce's death. As it was hard to see a sufficient number to replace them being recruited at all quickly, there appeared to be a very real prospect of a steep fall in income with a consequent need to cut back activities drastically.

While there was undoubtedly an element of truth in this, however, Pearsall was guilty of reading into the figures what he had been expecting to find in them. Had he only studied them more carefully, he would have discerned that there had been an abrupt drop in numbers which antedated Druce's death by some three or four years: it was the Depression, rather, that was the underlying factor. Forced to tighten their belts, people had been making the obvious first economy and pruning the number of organisations they subscribed to. The fact that the fall in membership levelled off just as the economy began to recover could perhaps be taken as confirming this[20] – but it is hard to be sure, as by then the Society itself had caught its second wind.

All the same Pearsall must be given much credit for the invigorating way in which he took the helm at a time when the Society would otherwise have drifted. In addition he provided it with some extra canvas to increase its momentum: apart from the strong lead he gave in the establishing of the new constitutional framework – in its broad outlines the one on which the Society has been based ever since – several valuable innovations that were introduced at this time turn out to have been his. It was he, for instance, who arranged in July 1933 the first field excursion in what was to develop from the next year into a regular annual programme (this time, unlike in the 1920s, unambiguously under the Society's official aegis).[21] It was at his insistence that the Reports went back to the old practice of listing members' addresses (a maddening omission of Druce's, about which Pearsall had remonstrated time after time with no success).[22] It was his idea, too, to widen and deepen the scope of the noticing of recent publications which Druce had made one of the most useful features of the Reports. To this end he invited two of the General Committee's professionals, Wilmott and Gilmour, to institute the annual Abstracts from Literature which has since matured into one of the Society's most keenly-appreciated services, winning it many extra members from overseas in particular.

By no means all the innovations that distinguished Pearsall's term of office came solely from him, however. Mrs Foggitt, for one, was responsible for initiating in 1932 an annual Conversazione in late autumn, the forerunner of today's Exhibition Meeting, on the model of, and for a long time in conjunction with, the Wild Flower Society's Tea-party of which for some years already she had been acting as the organiser. These occasions quickly proved extremely popular (about 140 members of the BEC alone were attracted to the second)[23] and were necessarily held in a capacious room at the Great Central Hotel in London. In 1932, similarly, another of the Society's stalwarts, Patrick Martin Hall, laid the foundations for the much more incisive methods of handling plant records that came into use in the years that followed. Sensibly setting aside Druce's *Comital Flora* as irremediably untrustworthy, he extracted with great labour all the records that had appeared in the Reports back to 1926, checked these for novelty against the base-line of *Topographical Botany* and all its various supplements and then filed the results on two sets of cards, the one arranged by vice-counties to serve as an index geographically, the other by species in systematic order. After that it was at last possible once again to state with confidence whether a record sent in was genuinely new for the vice-county in question or not. This core part of the BEC's work was thus put back on a properly cumulative footing.

That was not the only way in which the tidy-minded Hall, partner in a Portsmouth firm of chartered surveyors and a specialist from his schooldays in dactylorchids, placed the Society in his long-term debt. An arguably even greater service was his role as its chief lawmaker. Until the Annual General Meeting of 1935 the Society had somehow contrived to stumble along without any coherent set of rules, even omitting to fix terms of office for those elected to the General Committee. Now at last a detailed written constitution was hammered out, laid before the membership and, subject to a few amendments, voted through. The more diehard of the Druceans, nevertheless, regarded it with deep suspicion, seeing it as a device for setting in hard concrete the very different BEC that they found emerging. "We can but put up a fight, for the sake of the past – even if it is a losing one," wrote a resigned Mrs Foggitt, who thought the way this new constitution was "foisted" on the members was frankly "a disgrace".[24] Even so those who shared her outlook must surely have been delighted by the inclusion of Rule 2 (iv), for after a good deal of wrangling and the resignation of one particularly heated opponent, the conservation of the British flora had at last been proclaimed one of the Society's official objectives. At last, too, even more of a milestone in its way, membership was automatically cancelled for those who fell more than two years behind with their subscriptions.

Another, less noticed provision laid down that "The Committee shall prepare from time to time a list of Referees to whom it is recommended that critical groups of plants should be submitted for naming." Such a list in due course made its appearance as a supplement to the 1937 Report, in which it was announced as comprising "all those botanists who are contributing to the New Students' British Flora, now in course of preparation". This was a reference to the latest, and most promising, in the increasingly lengthy series of attempts to produce an up-to-date, truly comprehensive field textbook. In the absence of any solid competitor, an inadequately modernised edition of Bentham's 1858 *Handbook* continued to dominate this market – and thereby persisted in foisting on the unwitting its misleadingly broad treatments of many of the species. Now it was the turn of the Oxford University Press to try to improve on the efforts of its Cambridge counterpart of some fifteen years earlier. To this end three recognised authorities had been signed up as joint editors: Wilmott of the British Museum (Natural History), Gilmour of Kew and, to supply an ecological dimension, Oxford's Tansley. Though not a

BEC venture officially, as the principal body concerned with the advancement of knowledge on the subject the Society could not help but be closely involved: those who supplied it in the main with its scientific momentum were those to whom the project necessarily looked for its specialist contributors. Unfortunately, though, once again the initial drive and enthusiasm proved unsustainable, and even by 1939 there were still only two accounts in manuscript. For a couple of decades more, in consequence, the study which the Society embodied had to cope as best it could without this crucial centrepiece.

A side-effect of that failure was the similar non-appearance of a long-planned updating of the standard nomenclature in use. This had started out as yet another of Hall's initiatives. For just as he had been irritated past bearing by the sloppiness of the Society's record-logging system, so the many eccentricities of Druce's *British Plant List* of 1928 had driven him to take early action. Towards the end of 1934, accordingly, he had begun preparing for the Society a revised edition of that work. No more was to be heard of it, however; and as it was completely out of character for Hall not to persist with a task once undertaken, it must be presumed that it was laid aside deliberately once it became clear that Wilmott would be conducting a much more comprehensive overhaul of the current nomenclature as part of his work on the new national Flora. The subsequent collapse of the Flora thus left this second glaring lacuna that would also remain unfilled until the 1950s.[25]

Hall's varied effectiveness in these years suggests that he would have been a happier choice as Secretary than Pearsall. He was a man of greater all-round ability and easier to work with. Moreover, as one of Druce's twelve selected apostles, he was just as much of a 'bridge' figure – and at the opposite, more forward-facing end of the bridge to Pearsall. On the other hand, unlike Pearsall, he had a profession to attend to, so perhaps could not have handled the full range of the Secretarial duties.

At any rate, when Pearsall's death occurred, very unexpectedly, in August 1936, after he had completed four and a half years in office, Hall consented to take over as his successor only in part. The time had come, the General Committee acknowledged, for the hiving-off from the Secretaryship of the onerous editorial part of the work. The quietly efficient Chapple was accordingly promoted into a narrower Secretarial chair and Hall moved into the newly-created office of Editor. It must have proved a powerful partnership; but of the next few years, unfortunately, we know much less than we should, due to the incomplete survival of the various sets of minutes relating to the period (not that these are particularly informative documents even at the best of times). It is evident, though, that Hall lost no time in instituting improvements in the publishing of plant records in the Reports, a feature which he had been regarding for years with increasing concern and frustration. Two of his first steps were to try to standardise the form in which the records were sent in and to seek to ensure that voucher specimens were deposited in public herbaria, though it was not till some years later that the Editorial Sub-committee could be won round to the decision that a record could be rejected if no voucher could be produced. The publication of the Reports themselves was also speeded up, in the conviction that these lost a good part of their usefulness unless they were available for digesting around the start of the new fieldwork season. Furthermore, to consolidate his own earlier labours in bringing accuracy and comprehensiveness to the official register of what was on record for each of the vice-counties, he began work in 1938 on a thoroughly expurgated edition of Druce's *Comital Flora*, for this, despite all its known defects, was now in so many hands that it seemed sensible to base any new work upon it rather than proclaim a third edition of *Topographical Botany*. It was no doubt on Hall's initiative, again, that in 1938, also, the newly-founded Systematics Association was asked by the Society to call together a conference of biologists with a view to defining more precisely the boundaries of the vice-counties. The setting up of a joint committee of the two bodies was the practical outcome and the work of this was eventually to bear fruit in J. E. Dandy's definitive monograph of 1969, published by the Ray Society.

By 1939 Hall had so successfully chivvied the Society back along this more scientific path that at the Annual General Meeting that year it was decided to introduce the innovation of a lecture. This was a noteworthy step, for it was the first to be given officially for nearly ninety years. Appropriately, the speaker on that occasion was Hall himself, his topic being the one on which he had become the recently-revealed expert: the British species of *Utricularia*.

But it had been a long, uphill struggle to propel the BEC even as far as that. All through the 1930s the more dedicated of the Druceans had put up a vigorous resistance to the gradually strengthening power of those who wished it to become a higher-grade body. In the face of what they felt as a strangling constriction imposed by the drearily over-scientific, of whom the professionals were seen as forming merely one extreme wing, they strove to preserve the anarchical free-and-easiness of earlier years. Mrs Foggitt was the prime exponent of this ethos, preferring to regard the organising of the annual Conversazione as a matter for herself alone and refusing to acknowledge that she was responsible to a committee. Not for nothing had she been a pioneer aviator and balloonist.[26] In the end this behaviour became intolerable and she had to be dethroned. Pearsall, similarly, bowed under the yoke of the new Publication Sub-committee only after a series of angry protests addressed to him by Wilmott.[27] Even as late as 1939 the General Committee was still seen as divided into two opposing factions and Wilmott was talking of standing for re-election to it in order to "get the Society properly going".[28]

The break in the archival record in the latter part of the 1930s is fittingly symbolic. For the coming of the Second World War did more than enforce, as it did for most societies, an effective suspension of activities: it was to deprive the BEC of its Secretary for the whole of the duration and to lead to the premature deaths of both of the other two pre-war officers who remained. Hall, giving of his utmost as usual, overdid his Home Guard duties, suffered a breakdown in health and died in August 1941 at the age of only forty-seven. A few months earlier Francis Druce[29] (no known relation of the other), who had succeeded to the Treasurership around the same time that Hall had become Editor, was killed by a bomb which also destroyed his flat. All of the Society's records had been placed in his care and from the fire that raged not a trace of any of them remained. It was 1864 all over again.

But a clearly-defined period had in any case come to its natural close by then. The post-Druce transition had effectively been completed and it was time for a fresh team to take over and to broaden and deepen what had been started in those ten innovative years. For although the achievements of field botany in that period hardly look impressive when compared with those in British ornithology or archaeology, the recovery that the corporate study of the country's higher plants had made from its previous arrested development has indeed been marked. Although it was entering the 1940s a full generation behind those counterpart subjects in organisational reach and imaginativeness, at least it now saw where it wanted to go and had acquired the lineaments, if not yet the stature, of a national society of the kind it ought long since to have possessed.

Though not without much strain and even strife, a grand re-integration had taken place. The energies going into the study were ceasing to pull against one another at last. There was now but a single focus where previously there had been several.

The chief remaining obstacle to that unity, the existence of the Watson Club, had been removed in 1934. The periodic suggestions over the years that the two clubs should merge had predictably revived on Druce's death and, now that three-quarters of the Watson's members belonged to the other one too, the arguments in favour of such a move appeared to have greater force than ever. Nevertheless the deeply-embedded hostility that the leading Watson people, Pugsley above all, felt towards the BEC continued to stand in the way of this.[30] In the end, as a gesture of conciliation, the BEC, taking a deep breath, elected Pugsley to a vacancy on its General Committee. Not long after, bowing to the desire of their now-ageing officers to step down at last and accepting that a falling-off in support and the high cost of printing made it financially impossible to continue, the Watson Club voted itself into extinction.[31] Yet even then the long dream of an amalgamation still proved elusive: a majority of the members, egged on, it is said, by the ever-unforgiving Pugsley, opted instead for outright dissolution, seeing this as the "easier and more dignified" course[32] and as still leaving open the possibility that the Club might be restarted at a later date. This was a pity, because the name a new, joint body would have adopted, had Pearsall had his way, would have been 'The Botanical Society of the British Isles'.[33] The future already lay in blueprint down even to that detail.

PART THREE

*The Botanical Society
of the British Isles*

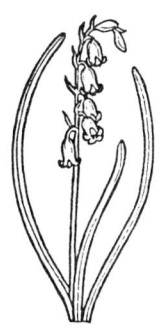

CHAPTER 11

The Great Efflorescence

On October 22nd, 1941 an emergency meeting of a sadly depleted General Committee took place in the Linnean Society's rooms. Its agenda was short: to consider what was to be done in the light of the loss now of every single officer and of all record of the finances.

Replacing the officers was much the lesser of the problems. Wilmott, dependably on hand at the British Museum (Natural History) – for he was over age for war service – proved willing to take on the Secretaryship *pro tem*. Along with E. C. Wallace, he agreed to act as Joint Editor too (though within a matter of months the rest of this second office descended on him as well, when Wallace was called up into the Royal Air Force). Highly experienced by now in the BEC's affairs, Wilmott's appointment must have been a great reassurance. But more reassuring still, though probably few realised it at the time, was the securing for the more immediately crucial role of Treasurer of Job Edward Lousley.[1]

More than to any other single person the post-war transformation of the BEC into the enormously more substantial BSBI was to be due to Lousley. At this time of assuming office he had been a member, and a notably active one, for fourteen years, having joined both this and the Watson Club at the age of twenty, ahead of his former schoolfellow, Wallace. He was to serve successively as Treasurer and General Secretary for the combined total of fifteen years, with a term as President and two separate spells as Vice-President after that. Throughout that period, which extended down to 1973, just three years short of his sudden death, he was also continuously secretary, chairman or an ordinary member of one or more committees. This record was the more remarkable in view of the fact that he followed a full-time non-botanical career in Barclays Bank, to which, unusually for a leading amateur, he was no less full-heartedly committed. In that respect he was a throw-back to the equally devoted and unwearying Bailey. Like Bailey, too, he was a man of great natural administrative gifts, which, given other circumstances, would surely have taken him to high levels in the corridors of Whitehall.

At the same time, in a surprising number of ways he also resembled Druce, though the two in other respects could hardly have been more different. Both of them were in large degree self-educated (Lousley's schooling, few were aware, had never extended to Latin); both earned themselves early reputations for uppishness from the staider of their botanical seniors; both, equally, were prey to an insecurity which rendered them incapable of taking criticism. Both, again, were life-long, incurable collectors; both had very deep and extensive field knowledge, built up the hard way in straitened youth; both were ever keen to give help and encouragement to beginners; and both came to identify themselves with the Society totally, resulting in the release of torrential energies in its service. In contrast to Druce, though, Lousley was not a genial person: austere even, and starkly forthright in criticising others, despite his own personal touchiness. In these traits, yet again, he reminds one of no one so much as Watson. Watsonian, too, was perhaps his one most important characteristic of all: an approach that was effectively that of a professional.[2] There was a rigour that he brought to his scientific work which set a new high standard for the amateur but which had the effect, in turn, of placing him to some extent above and apart. Because of this, he was the ideal person to run the Society at a period when it needed above all else to gain the respect, and even better the allegiance, of the many exponents of the New Systematics who were beginning to emerge in the universities.

Wilmott's background was not all that dissimilar; he was like Lousley, too, in his orderliness and rigour. Their respective choices of genera on which to cut their taxonomic teeth – Oraches (*Atriplex*) in the one case, Docks (*Rumex*) in the other – hint, furthermore, at a common streak of puritanism in their make-ups. But where Lousley was touchy, Wilmott was volcanic: unpredictable, even at times a little unbalanced. But his ferociousness on the whole was reerved for his peers: to novices, and to anyone else in an unthreatening, supplicant role, he could be tirelessly kind and patient. Burly and muscular, it was as if his need to dominate arose from his physique. The one, unlikely sport that he played, table tennis, he threw himself into with a demonic intensity and, it was said, he could never bear to lose. Almost as unlikely, he was a highly accomplished pianist. Indeed he was one of those people whose level of achievement in life tends to be lowered by the very diversity of their gifts; certainly, his published output was in no way commensurate with his depth of knowledge and taxonomic acumen. Too easily tempted by every new hare, he was the victim of his own creative restlessness.

He and Lousley worked well enough together – for both wished to see the same kind of society develop – but they were never to be close. For a long time, in any case, their paths had little cause to cross. Most of the BEC's activities had been suspended for the duration, so that Wilmott's duties were

G. C. Druce examining the Loddon Pondweed in Dorset 1931.
Described by Fryer as a supposed new species *Potamogeton drucei*,
this is now regarded as *P. nodosus*, already known elsewhere in Western Europe.

Potamogeton nodosus Poir.   Loddon Pondweed
(syn. *P. drucei* Fryer.)

A, flowering stem; B, flower—one perianth-segment and one stamen removed; C, perianth-segment and stamen; D, immature fruit and section of same—fruit of wild-collected British specimens not seen; E, F, submerged leaves.
Perianth green; anthers yellow. Leaves bright green.

Loddon Pondweed.
From a drawing by Stella Ross-Craig in *Drawings of British Plants* (1973).

largely on the editorial front, and even there much reduced, for the normal flow of material had shrunk to a trickle and the Reports were necessarily scantier and fewer. To Lousley, by contrast, fell the incomparably more laborious task of entirely reconstructing the finances. This proved very much more difficult than expected, because the last published membership list turned out to be seriously inaccurate. Even after two circulars in succession and month after month of intricate detective work (for the War had caused many to change their addresses more than once) there were still an irreducible 44 out of the putative 446 who remained elusive. Of the rest, a goodly number claimed to be unaware that their names either were or ever had been on the books;[3] and there were four who stated that "by arrangement with the late Dr Druce" they had been excused ever paying a subscription and whom it was thought best to make Corresponding Members, therefore, in order to regularise the position. A further 30 proved to be deceased. Thus for one reason or another one member in every four had to be written off. The total number of subscriptions remaining yielded only some £150 a year at the then existing rate of ten shillings, an income quite insufficient had it been necessary to sustain the pre-War level of activities.[4] But luckily it was not; and indeed in recognition of the great reduction in what was being provided in return for the subscription, it was felt that this must be halved until at least the Reports began to appear regularly again. The Society was able to keep alive only by going into an extended hibernation.

The first sign of a reawakening came in January 1944, with a return to the old rate of subscription. The justification for this was to have been the issuing of a bumper, 'catching-up' Report for both that year and the previous one combined; but once again the Society proved to be out of favour with the gods: the Editor's health was not up to the task and it was not until 1946, to general embarrassment, that the volume made its eventual appearance.

By that time most of the other pre-War activities had restarted, one by one. The first to do so, ironically, had been the Exchange – ironically because for a long time this had been steadily losing support and its following now sagged abruptly further, from eighteen contributors to a mere twelve on average, reflecting the passing out of fashion of the forming of private herbaria. Such a sorry decline gave heart to those who felt that the carrying-on of this service gave a damagingly wrong impression of what the BEC now mainly stood for and undermined its reputation. Even so it was not for another ten years, in 1955, that any proposal was made for its abolition. By then the climate had become more generally hostile still and in the event, when the case for retaining it was pressed to a vote, only a solitary hand was raised in its defence.[5]

Already, though, eight years before, in 1947, the 'Botanical Exchange Club' itself had been symbolically done away with. Many members, perhaps even most, had been dissatisfied with that name for a very long time, but the wish for a change had been outweighed by habit. Now at last the ending of the War provided the necessary stimulus. In October 1945 a new sub-committee had been set up – in effect an inner core of the General Committee – consisting of Lousley, Wilmott, Pugsley, Gilmour, E. Milne-Redhead and Miss M. S. Campbell as Secretary, to draw up plans for the expansion that was envisaged and for the future lines of development. One of its first actions was to send out a questionnaire to establish the popularity of various existing and potential activities and to canvass opinions on a number of mooted changes. A change of name was foremost among these. Several suggestions to this end emerged, but not all were found acceptable: 'The British Flora Society' and 'The Flora of Britain Society', in particular, were speedily rejected once the acronyms that these would give rise to were realised. The 'Botanical Society of the British Isles', a strong favourite up till then, the sub-committee firmly set aside as inappropriate, in view of the fact that not all categories of botanist were catered for nor the whole of the plant kingdom covered. Instead it came out in favour of 'The British Society of Field Botanists' (a suggestion of Miss Campbell's). The General Committee, however, did not take to that at all and, following a postal ballot in which forty per cent of the membership voted, 'The Botanical Society of the British Isles' was settled upon by an overwhelming majority. In October 1947 that name was formally adopted.

To accompany this outward change of clothes, the opportunity was used to update the wardrobe more generally. The General Committee was renamed the 'Council' and, instead of continuing to hold leisurely all-day meetings, began to have brisker ones in the evening, thus suiting the convenience of those amateur members who had full-time occupations. Other changes were the embarking upon a series of biennial two-day conferences "with some trepidation";[6] the introduction of Junior Membership for those under twenty-one;[7] the appointment of a chain of vice-county Recorders to complement the tasks of the already-existing Local Secretaries; and the extending of the panel of referees on critical groups to cover advice on various specialist topics as well.

Suitably coinciding with all this, a major change-over of officers also took place. After fifteen sagacious years as Chairman, Baker at last intimated his wish to step down, and in his stead Gilmour was elected. Soon afterwards, as part of the epidemic of renaming, the latter thus had the honour of becoming the BSBI's first 'President'. At the same time Chapple also found it necessary to go and handed over to Miss Campbell. He had duly come back after the War to his former post at Oxford and relieved Wilmott of the Secretaryship – or rather, as it became from 1946, the 'General Secretaryship' (for Wilmott carried on as Secretary to the General Committee). Before long, however, the anomaly of his status had brought him under pressure from the University and led him to accept an opening that had been offered to him in publishing; this proved to require his being overseas for lengthy periods, so it was impracticable for him to carry on in office. Wilmott, in turn, had decided this was the point at which to resign as the senior of the two Editors and he gave way to E. F. Warburg, who was effectively Chapple's replacement at Oxford. Of the pre-existing team only Lousley thus remained, and he did so only at the price of the appointment of an Assistant Treasurer in the person of E. L. Swann, to help him carry the burden.

This last change was to prove of critical importance politically, for it freed Lousley sufficiently to allow him to play the leading part in the two years that followed in resisting the attempts at domination by Miss Campbell and Wilmott. The appointment of an Assistant Treasurer was also a noteworthy development in that it was the first post in what was to be a further tier of management made necessary by the ever-expanding workloads of the principal officers. It was a solution that was to be increasingly resorted to henceforward, as the preferred alternative to departing from a regime that was now wholly voluntary by taking on paid assistance.

May ('Maybud') Sherwood Campbell, the new General Secretary, was temperamentally out of sympathy with this policy from the first. With that fondness for grand gestures characteristic of the concert singer that she had been till lately it was unlikely that she would ever have taken comfortably to running a body so wedded to austerity. With hindsight it can be seen that the choice of her for this pivotal office was for that reason a mistake. Yet it seemed at the time an obvious and sensible appointment: she was based in London, she was free of the ties of a family or career, she had shown herself a good organiser in her running of the Society's field meetings programme over a lengthy period, she was a capable field botanist and, although (unlike any previous holder of the office, except perhaps Bailey) her scientific standing was but slight, she was a close friend of Wilmott and so had him to turn to for all the help in that direction she might need. Some may even have supposed indeed that she was Wilmott's marionette. Their partnership, however, was a much more equal one than that; they functioned rather as British field botany's Ferdinand and Isabella, reigning over it jointly. It was a reign, though, that must have been expected to be tempestuous, for both of them were people who liked to have their own way and did not shrink from tantrums or bluster. Already, indeed, just before the War, Miss Campbell had even gone so far as to resign from the Society for a time, after the General Committee had failed to act on some adjustments to the field meeting guidelines that she had stipulated.[8] On the other hand, it must be said, she was by no means alone in those pre-War days in displaying difficult behaviour: a generation still accustomed to flounce out of grocers' shops if not served promptly had far less patience with the frustrations of democratic procedures than those who have followed since.

The new Editor, by contrast, was a much more auspicious figure. Edmund Frederic ('Heff') Warburg was the first-ever university teacher to be appointed to one of the Society's executive offices.[9] That in itself was an important milestone. Even more important, though, he had been picked out for this key position expressly because of his ability to represent the 'new' taxonomy, that amalgam of novel, much more incisive techniques that was promising to revolutionise the traditional approach. Already he had brilliantly collaborated with Wilmott in sorting out, though he had yet to publish his work on, those long-recalcitrant groups, the Whitebeams (*Sorbus*) and the Lady's Mantles (*Alchemilla*), and had moved on from those to the even more recalcitrant Birches (*Betula*) and Eyebrights (*Euphrasia*). Afflicted in these earlier years with an agonising shyness, he was not a person who shone in committees; but his strengths were well enough known to those who mattered. Moreover, in a lengthy letter he had written in 1946 around the time of his joining in response to the recent questionnaire, he had manifested a keen concern that the Society's Reports should be deepened in their scope and so take the place of the *Journal of Botany* (which had foundered in 1942) as the main outlet for papers on the systematics of the British flora. The new publication that he envisaged "might well include a number of features of the present Report", he had suggested, "leaving the remainder ... to be dealt with in an abbreviated version". It "should be such as to attract foreign botanists and institutions and botanists in this country less closely connected with systematic botany", and on the basis of three or four issues a year, he supposed, it "would surely in due course be self-supporting".[10]

Within three years that proposal had become reality. The Council shared Warburg's ambition entirely (though hardly his optimistic economics) and was happy to entrust to him the necessary planning and editing, reserving to itself only the choosing of the journal's title. This, it was decided, by nine votes to two,[11] was to be *Watsonia*, in commemoration of the man whose particular set of interests the Society faithfully embodied and to whom it substantially owed its very survival. The inaugural number eventually appeared, much later than intended, in the first days of 1949 and had a very encouraging reception. Appropriately, the main item in this was a definitive overhaul, by S. M. Walters, of that hitherto much misinterpreted group, the Lady's Mantles, on which Warburg himself had done invaluable groundwork already. In the next part there came the first in a series of outstanding papers by D. P. Young elucidating at last the more baffling of the autogamous Helleborines (*Epipactis*). After that came one announcement after another of notable additions to the flora of the British Isles: first the overlooked Parsley Piert segregate *Aphanes microcarpa*, then a further Marsh Orchid from Ireland, *Dactylorhiza cruenta* – subsequently to be considered only a subspecies of *D. incarnata* – then a Horsetail from Lincolnshire, *Equisetum ramosissimum*, then the rediscovery of the long-lost Guernsey grass, *Milium scabrum*. It was a promising start. Even so, many journals begin with high promise, but fail to maintain their quality all too quickly. Thankfully in this case good material continued to flow in, and before long it was safe to claim that the publication was established. Warburg was to continue to act as its sole or joint editor for altogether fourteen years, by which time he had proved his contention many times over that there had been an unfilled niche in the botanical literature. Ever since its founding, indeed, *Watsonia* has served as the principal outlet for papers on the taxonomy, and especially the experimental taxonomy, of the vascular plants of the British Isles, far outstripping in its near-monopoly of this field all competitors (of which the *New Phytologist* and the *Botanical Journal of the Linnean Society* have been perhaps the most significant).

To complement the new journal, Warburg's further suggestion of hiving off the purely domestic matter the Report had hitherto contained was followed up by publishing it in a *Year Book*. After only five issues, however, the Society's vaulting ambition overtook this development in turn and in 1954 it was expanded into a twice-yearly *Proceedings*, which otherwise broadly paralleled *Watsonia* in pattern but carried papers of a more traditional, less dauntingly technical kind; it also contained an increasingly lengthy and more valuable section devoted to abstracts from the world-wide literature. Edited for the first twelve of its fifteen years' existence by D. H. Kent and after him by E. F. Greenwood, this supplementary journal ideally solved the problem that the Society now had of needing to satisfy two largely separate levels of readership. This was all the more necessary after 1953, when, in order to attract a predicted greater flow of papers from the increasing number of research students in the universities, *Watsonia* acquired a larger page size to allow for more spacious diagrams and tables and became considerably more technical than previously. It also became much costlier to produce, for the by now unacceptably antiquated Buncle's were at last forsaken for a much more expensive firm of printers.

This heavy outlay on periodicals, however, was not the whole of the amount devoted to publications at this period. In addition, it was felt to be important to bring out the proceedings of the Society's conferences as a series of printed volumes. This entailed further big expenditure every other year. Annual applications to the Royal Society for assistance from the Scientific Publications Grant-in-aid Fund successfully attracted £800 in all in the course of the first five years, which besides the financial relief it afforded was heartening recognition of the standard that was now being achieved. But such sums could never be counted on, and there was the further likelihood that printing costs would continue to climb at a disproportionately rapid rate; so in deciding to embark on so bold a programme, the Society had clearly taken a very considerable gamble. Lousley, however, was confident that, with all it now had to offer, it was capable of doubling its membership; and although some of the Council were incredulous, he succeeded in carrying the majority with him on this.[12] If anything approaching such an increase was to be obtained, it was obvious that a special effort would need to be put into promotion. To this end accordingly a small Advertising Committee, made up of Lousley, Chapple and R. A. Graham, was appointed in the spring of 1950.

A myth has somehow taken hold within the Society, albeit understandably, that the main post-War surge in its growth was the consequence of the later Maps Scheme. The truth is otherwise: it was the result of the extraordinarily intensive recruitment drive that was mounted in the early 1950s, before that Scheme had so much as started. And for a good deal of that surge the simple, humdrum toil of that historic, three-man committee, addressing envelopes to every likely non-member they could think of and then posting off in these a copy of the Prospectus, was unquestionably responsible.[13] In 1950 alone no fewer than 706 individuals and 862 institutions were approached through that means. As a supplement, advertising space was taken in appropriate journals and many other individual members, caught up in the general enthusiasm, helped with their own, often local soliciting. In some parts of the country, in the North of England especially, there was not just apathy but actual hostility to overcome: the old, even pre-Drucean image of an over-collecting coterie still stubbornly lingered on. To these benighted regions the publications of the reborn BSBI came as a revelation, and their botanists joined in droves. As a result of all these efforts, from the second half of 1950 the membership figures started to rise dramatically. As against a mere 12 in 1949 the net gain in that next year proved to be a highly gratifying 110. In the two succeeding years the hauls were 79 and 97. Altogether, in the course of just that short

Miss M. S. Campbell.

A. J. Wilmott (left) and Francis Druce at Arisaig, 1935.

period alone, coinciding with the life of the Advertising Committee, the Society swelled by fifty-seven per cent, from 498 to 784. By the time the next three years were out the total stood at 1038 – and Lousley had achieved his target. Even so there could be no letting up yet: the Society still needed to grow as fast as it could if its desires were not to outreach its purse.

Such a rapid inrush of further members on top of so many new and expanded activities – and, not least, so many more meetings of committees – naturally placed a great strain on the principal officers. Had their efforts not been rewarded so abundantly and so visibly, several might well have found it too much. In the event the casualties were limited just to one and that comparatively early on.

The casualty was Miss Campbell, though her departure was not of her wishing. Although she had played a leading part in the planning of the multifarious new developments, it is clear that she had greatly underestimated the demands that the resulting extra work would place on her energies and leisure. At the same time, it must be said in her defence that no one had realised that the General Secretary now needed to be much more of a co-ordinator than previously and so ought to shed several of the duties that had become attached to the office (in some cases because there was no one else they obviously belonged to), in order to give adequate attention to that priority role. It was a classic instance of administrative structure having fallen out of step with events.

It would also seem that she was allowed to accept the office without an accurate idea of what it entailed. She had been on the Council for only just over two years and her predecessor, Chapple, had run things from Oxford largely unobserved. Such impressions as she may have obtained from Wilmott are likely to have been coloured by his own highly abnormal wartime experience.

Unwisely, as it turned out, she chose to retain at first her long-standing responsibility for the programme of field meetings, handing over to J. G. Dony just the indoor part of that portfolio (which included the very onerous conferences). The respective titles of 'Field Secretary' and 'Meetings Secretary' were thereupon introduced to reflect this dividing-up of that side of the Society's work. In addition, in view of the fact that it was essentially an overall report on the Society's activities, the General Secretaryship also landed her with the editing of the *Year Book*. Little noticed, too, was that old survival from Druce's days, the standing invitation to members to send in to the Society's chief officer any non-critical specimens they wished to have named, a service which was believed to be responsible for attracting and holding many new recruits. A few of those who had recently joined took advantage of this service to the point of abuse, with the result that it was later devolved on to a further set of local officers specially created for the purpose, confusingly termed 'Referees'. But that was after Miss Campbell's time: this was a duty that brought her into touch with the less knowledgeable members and one which on that account she relished more particularly; and here her British Museum (Natural History) association came into its own.

Sensing that she was in a strong position, she had accepted nomination on a year's trial and on condition that she was granted "expenses on a fairly generous scale for secretarial assistance with correspondence (short-hand typist) etc." It was also agreed that she could purchase a duplicator. Lousley, who by chance had missed that particular Council meeting, was horrified on learning of this. He had been anxious that Chapple's successor should be someone of scientific weight, ideally a professional. Not only was he disappointed in this, but in his view the Council had conceded ruinous terms. "It involves a fundamental change in policy with serious effect on our finances," he complained: "In recent years the Officers[14] have kept expenses down to the very minimum – even using their time to save the Society money in such ways as re-using old envelopes." At the express request of the Officers, he pointed out, there had been a decision against the payment of any honoraria,[15] yet "even under these conditions we have only paid our way".[16]

Although ostensibly he advanced his objections on the ground of financial prudence, in reality they went much deeper than this. A society such as the BSBI, he felt as a fervent matter of principle, should be run on the most sparing basis possible and the subscription accordingly kept as low as it could be so that there should be no shutting out of the impecunious (as he himself had once been). It was the argument which had torn apart the old Botanical Society of London surfacing afresh. Lousley, with his background of youth-hostel penuriousness, personified the one point of view; Miss Campbell, with her background of comfortable affluence (which led her, for example, to charge up taxis unthinkingly), equally personified the other. At the same time, as well as a conflict between two rival ethics, it was also a contest between two rival but equally respectable notions as to how a society of such a size can best be run. Many comparable bodies, as Miss Campbell was not slow to point out, operate on the 'chief executive' principle, attaching prime importance to securing the services of an able, highly-committed person and willingly meeting the price that this normally implies: if not a good salary, then at least comfortable expenses and sufficient paid assistance to liberate them from the lowlier clerical routines. The approach is the commercial one of 'buy the best and keep them happy'. The opposing philosophy, which is that of the friendly societies and collectives, prefers to put its trust in voluntarism exclusively, despite the greater uncertainties inherent in that course. Because in a regime of such a kind resentments tend to arise if anyone is treated preferentially, workloads and responsibilities have to be apportioned with a carefully measured evenness and no one individual can be accorded an outright supremacy. This latter philosophy is viable only so long as the collective workload remains not too heavy for it to continue to be subdivided among a flow of ready volunteers. Once so ideal a system has been adopted, however, its attractions are such that there are few who will want to move away from it until this has become absolutely essential, not least because of the sudden heavy impact on the individual pocket as a result. The great problem always is deciding when precisely that point of surrender has been reached and the alternative philosophy must be embraced, in whole or in part. It is a problem with which the BSBI has had to grapple all through the post-War years, and Miss Campbell was not to be the only General Secretary to force the Council to confront the dilemma seriously.

Initially, Miss Campbell had her way. The Council agreed that she should claim expenses up to a maximum of £20 a year, which was enough to enable her to employ the proposed typist for a few hours on average each week. In addition, two months after entering office, she was given the services of an Assistant Secretary, in the very efficient person of W. R. Price. On to him fell the main burden of editing the *Year Book*.

But before long the full extent of what she had taken on was all too overwhelmingly brought home to her and she began to agitate for an honorarium (as a mark of recognition for her services, for she was possessed of an ample private fortune). When, after a year's deliberations, this was conclusively denied her, as being a class of expenditure the Society's existing finances were unable to stretch to, she returned to the attack in February 1950, with an explicit ultimatum: the subscription, at that time still a very modest guinea, must be raised, so that the remuneration she felt entitled to could be afforded, and two further people must be appointed to assist her; otherwise she would have no alternative but to resign. At the long and fraught meeting that followed, which has remained vividly in the memory ever since of all who were present, the Council felt able to fall in with her requests only to the extent of appointing a single assistant: an Advertising Manager, a temporary office which Chapple proved willing to shoulder. Her resignation was accordingly tendered and, to her dismay, accepted. And as if to deepen that wound, Lousley, her arch-antagonist, thereupon agreed to take over in her stead.

It was as well that she had overplayed her hand, for the friction with those she had to work with would only have intensified. Moreover Wilmott, all along her mainstay, had died just a short time before, very suddenly, and without his presence the office must surely have lost for her much of its savour. Incomparably weighty as an ally, he might well have carried the day for her had he only lived to speak on her behalf. Deprived of his support now, she could only continue to lose battles. It was an appropriate time to go.

There was more to it, though, than just one individual's defeat. In a subtle kind of way her departure symbolised the fading of the pre-war BEC. For, as the more perceptive must have sensed, it had been a trial of strength at a deeper level than a mere clash between two schools of thought about how a learned society should be run. At bottom two sharply different worlds had been at odds. The kind of BSBI that was now emerging was incompatible with the outlook and atmosphere that Miss Campbell had cultivated and stood for. One way or another there would soon have had to have been a change of regime.

CHAPTER 12

Full Steam Ahead

No one who was a member through the early 1950s will easily forget the excitement of those years. Many a learned society has the luck to fall upon a phase when its aims are all being realised and recruitment is rising fast. In such golden periods the very fact of feeling on the move imparts its own dynamic. So it was for the BSBI when the new decade opened. There was an exhilarating sense of pounding ahead on all engines: almost every conceivable activity appeared to be under way and being performed with high efficiency.

In such a mood it is dangerously easy for a society to slip into overconfidence, to aim beyond its grasp and to embrace innovation excessively. To some extent this happened to the BSBI at this period. Proposals began to come in from groups of members who had not been associated with the planning of the first main cluster of developments and, just because these proposals emanated from the Society's grass-roots, they were treated with more reverence than perhaps they deserved. Spontaneous shows of concern from the periphery are always highly welcome to those who labour at the centre, but they are liable to be misread as currents of representative opinion to which there is a special obligation to react with indulgence. In the absence of any system of gauging general opinion accurately, those who have the task of charting a society's course are all too apt to clutch at anything that bears the appearance of a guiding responsiveness.

In 1952 a group of Cambridge members put forward two suggestions which had particularly far-reaching consequences. The first was that more should be done to foster an interest in field botany, and more specifically in what the Society had to offer, among young people of school age. This led to the setting up of a Junior Activities Committee, under the Secretaryship of A. W. Westrup, a schoolmaster turned technical college lecturer, which in the course of its thirteen years' existence organised a regular programme of meetings, including several of some length on the Continent, specially tailored to the interests of this age-group. In addition, schools and youth organisations were very extensively circularised. So very striking was the amount of effort put into this area that it was subsequently, in 1959, felt appropriate to promote the secretary of the Committee to the status of a full-scale Officer. Unfortunately, though, the early momentum did not persist; and eventually it was acknowledged that this side of the Society's activities was receiving disproportionate recognition and both office and Committee were quietly allowed to lapse.

The second 'Cambridge proposal' to bear sizeable fruit was that more should be done for those many members who were outside easy reach of London. This touched the Council and officers on a particularly sensitive nerve, for they were uncomfortably aware that the Society's London base had led to a marked south-eastern bias in the composition of committees. Though in large part this resulted from the strong representation on these of the British Museum (Natural History) and Kew, and though it could in any case be justified as no more than reflecting the fact that almost exactly a third of the membership had Home Counties addresses, there was all the same a strong wish to counter the geographical top-heaviness. The most obvious step to this end was to start holding regional equivalents of the April lecture meetings which had grown up as a surround to the AGM, but with exhibits and a field excursion added as extra bait, to make the occasions sufficiently substantial to draw in members from as wide a radius as possible. An initial experiment on these lines at Manchester in 1953 proved highly encouraging and it was decided to make a point of holding one such meeting annually, in a different part of the British Isles each time. In 1957 the trend was taken further and it was agreed to hold the AGM itself outside London in alternate years. Meanwhile, quite coincidentally, a proposal had come from Scottish members that problems caused up there by the parallel existence of the BSBI and the Botanical Society of Edinburgh could best be resolved by the forming of a joint committee to organise a common programme and co-ordinate field-work. Thus emerged in 1955 the Committee for the Study of the Scottish Flora, which was to last in the end for twenty-three years, for twenty of those under the Chairmanship of R. Mackechnie and for fifteen under the Secretaryship of B. W. Ribbons. Wales and Ireland subsequently followed this example, in 1962 and 1963 respectively, though these were to be wholly BSBI initiatives.

Towards the end of the 1950s this strong swing towards regionalism was taken to its logical extreme by amending the Society's constitution so that a regional element was built into the make-up of the Council itself. For this purpose the British Isles were divided up into seven constituencies, each of which took it in turn to host a Regional Meeting and in the course of this to elect a 'Regional Representative'. At the same time the opportunity was taken to streamline the network of local officers based on the vice-counties, which had been allowed to grow unrealistically elaborate. One whole tier, the Referees (for non-critical material), was abolished, their role being transferred to the Recorders; while the third tier, the Local Secretaries, was drastically compressed into just fifteen larger groupings, based on a series of newly-defined Society 'Districts', broadly equivalent to Watson's one-time groupings of vice-counties that he called 'provinces'.¹ The 'District Secretaries' responsible for these, however, failed to catch anyone's imagination. And indeed the new regional and local structure as a whole was too abstract in conception to stand a serious chance of functioning effectively. Except for the vice-county Recorders, with their straightforward bedrock place in the system, most of it remained a mere paper pyramid. Even the Regional Representatives turned out to be a mistake, for their elections tended to be hole-in-corner affairs compared with the sometimes intense competition for Council vacancies when the AGM was in London, which produced an inevitable feeling that the regional seats were soft ones, insufficiently 'earned'. After being persisted in with what seems now an excessive patience, most of the structure was accordingly dismantled in 1972, leaving only the ever-indispensable Recorders and only Scotland, Wales and Ireland with their reserved seats of their own on the Council. It had proved to be another drive that had ended in a cul-de-sac.

Yet these were essentially side-developments, which took a long time to unfold. The Society in the meantime was fruitfully engrossed in matters that concerned it more centrally.

One of these was the safeguarding of the sites of rarer species. Although in the 1930s the BEC had forthrightly associated itself with the efforts of the Wild Plant Conervation Board (set up in 1931 by the Council for the Preservation of Rural England), not much had been done officially in this direction under its aegis. After the War, however, the sudden, major surge forward that the conservation movement had taken in Britain during the early 1940s had found its reflection within the Society in a heightened degree of attention devoted to this aspect. Local Secretaries were given the newly-emphasised role of alerting the Officers to habitat threats which developed in their areas, and a subscription was taken out with a press cuttings agency to improve awareness of public enquiries taking place, so that the Society could arrange to be represented at these if necessary.² Quick decisions were often needed in such cases and necessarily fell to the *ad hoc* committee of the Officers which had existed since 1932 to take action on urgent matters. In October 1948 Wilmott was co-opted to this to free the other members from the often very time-consuming work that proved to be involved, and within a year a special sub-committee had grown up for the purpose around his efforts. Known for a time as the Threats Committee (its meetings largely consisted of threats by the General Secretary to resign, said the cynical), this was rechristened the Conservation Committee at the end of 1950. In that same year regular liaison meetings with officials of the newly formed Nature Conservancy began to take place, the first in a long line of developments which helped to ensure that the Society's activities on this front formed an integral part of the overall research and campaigning nationally.[3]

Remarkably, Lousley proved happy to assume responsibility for this new committee on top of all his duties as General Secretary.[4] The latter he may indeed have considered almost a rest-cure by comparison with the Treasurership, which he had doubtless been relieved to pass on to his Assistant in that office, Swann. At the same time Dony acquired responsibility for the full range of meetings, the organising of the field ones having eventually been relinquished by Miss Campbell in April 1949. In recognition of this, four years later his title was changed to that of Meetings Secretary and the arrangement of the field programme was delegated to a separate Field Secretary, initially subordinate to him (at first O. Buckle, then from 1957 P. C. Hall). To complete the reshuffle, in 1953 Kent succeeded Price as the Assistant Secretary responsible for membership and the *Year Book* – an office which soon became a great deal more onerous when the latter swelled into *Proceedings*. With his multifarious labours behind the scenes, indeed, Kent was to become almost a shadow General Secretary, freeing Lousley to throw himself into conservation matters to a much greater extent than otherwise would have been possible. Finally, to help relieve the pressure all round, more especially on the conservation front, and to take over the minuting of the Council meetings, a further Assistant Secretary was created, in the person of Mrs B. Welch. It was an exceptionally strong team, certainly a stronger one than the Society had ever had previously.

The order of importance in which these offices ranked administratively only vaguely corresponded, of course, to the order of priority the membership at large attached to the various spheres of activity. As always, too, some officers were much more visible than others, the nature of their work bringing them into touch with numerous members in person. Of none was this truer than the Field Secretary, for the new holder of that office made a point of being present at most if not all of the meetings each year, in order to observe the arrangements and be on hand if anything untoward occurred. For sometimes unfortunate things did happen. Perhaps the most notorious was at the Taunton Meeting in 1949, when two members, desperate to make an unscheduled visit to nearby Brean Down, in effect commandeered the coach. Though their action was the result of a misunderstanding, Miss Campbell made an issue of it and insisted on a letter of reproof being sent to them by the President on behalf of the Council.[5] Ever since 1936 it had been the Society's declared policy not to direct field meetings to well-worked localities like that, "with their special and frequently very rare plants"; instead, they were designed to bring extra attention to bear on places or districts which appeared to have been inadequately studied or which afforded special opportunities for the exposition of a particular critical group. Since that time there had normally been at least one meeting yearly that extended to as long as a week, so that a party which was prepared to be assiduous had the chance to make an extremely decisive impact on an underworked area. In 1950, for example, two-thirds of the recorded vascular species of the Isle of Man were listed in the course of a six-day visit (many of them in new localities) and some twenty additional ones noted as well. Such feats were only possible thanks to the fact that in those early post-war years most meetings attracted several of the country's most knowledgeable field botanists, including a sprinkling of professionals. Private motoring had barely begun to recover and attendances were huge by comparison with today's. In 1947, it was calculated, over the year as a whole they accounted for a third of the entire membership. Large numbers made the hiring of a coach economic, and travel thus *en masse* was conducive to a pervasive sociability which somehow was lost when coaches later gave way to cavalcades of cars. One meeting, that based on Carlisle in 1949, has even been immortalised in literature, a chapter being devoted to it, anonymously, in Dorothea Eastwood's *Mirror of Flowers*.[6]

Another feature of the early 1950s, a measure of confidence in the Society's enduring prosperity, was the launching of work on two much-needed volumes of reference. One of these, the concern of a special sub-committee, eventually appeared nine years later, under the authorship of the infinitely painstaking J. E. Dandy. This was the *List of British Vascular Plants*, the product of a merger of three previously independent and competing initiatives and symbolically published by the Society and the British Museum (Natural History) jointly. 'Dandy', which is still in print, became at once the unchallenged successor to the obsolete *London Catalogue*, that long-running survivor of the ancestral Botanical Society of London, and it imposed a standard on the contemporary nomenclature which has effectively held sway ever since. The other volume, fortuitously also published in 1958, was *British Herbaria*, a series of indexes to the location within the British Isles of all the surviving collections that could be traced. The work in the main of D. H. Kent, it was published with the aid again of a Royal Society grant.

Far overshadowing all of these activities in magnitude and importance, however, was the great Distribution Maps Scheme, which was to culminate in the publishing of the *Atlas of the British Flora* in 1962.

The Maps Scheme had its birth in a carefully stage-managed conference in 1950, under the title of 'The Study of the Distribution of British Plants'. The planning of this had begun a full year before and the suggestion of the topic in the first place is said to have been Lousley's. With a greatly increased membership in prospect, the time was certainly ripe for a major co-operative enquiry of the kind that had long been familiar to ornithologists. Such an undertaking, it was perceived, would give the Society a tonic sense of united purpose, while encouraging a much higher proportion of members to direct their efforts into patently productive scientific work. In the process the Society could expect to gain greatly enhanced publicity and esteem, as a result of which its numbers were likely to be boosted helpfully further. What is more, mapping the country-wide distribution of the individual species of flowering plants and ferns had been a central preoccupation of the Society in an earlier phase of its existence: what was now in mind was the natural continuation, one stage further on, of Watson's successive compendia, *Cybele Britannica* and *Topographical Botany* – in effect, the ultimate destiny, the realisation in visual terms, of the Society's record-collecting efforts all through the years.

By a lucky accident of timing, the modern, comprehensive textbook that field botanists had been patiently awaiting for so many years had just then made its appearance and would be able to serve as the necessary standard for reference. Clapham, Tutin and Warburg's *Flora of the British Isles*, published in 1952 by Cambridge University Press, was an independent venture and had not involved the Society even to the tangential extent of its abortive predecessor in the 1930s – with which, in turn, the new Flora had no more link than a common inspirer in the person of Tansley.[7] Its advent at once made a very marked difference to the general taxonomic standard, enabling the average worker in the field to cover species and subspecies of which he had hitherto fought shy or even been quite oblivious. For the incoming wave of postgraduate students (and no less for their supervisors) it also pinpointed those groups on which their deeper investigations could most profitably be focussed.

At the 1950 Conference the various methods employed up till then for delineating distribution patterns were critically examined. It quickly became clear that the time had now come for British and Irish botanists to adopt a method of mapping that would permit the recording of fuller detail than was possible with the long-standing vice-county system. A description by

Professor A. W. Kloos, Jr. of the work in the Netherlands was particularly instructive in this connection. The upshot was a proposal by Professor A. R. Clapham, one of the country's foremost ecologists, that the launching of a comparable British Isles project should be looked into forthwith, with the 10 km. squares of the National Grid as the suggested area unit.[8] The proposal was carried with acclamation.

A special committee set to work very shortly afterwards to ponder ways and means, with Lousley in the chair and Clapham acting as Secretary – for what was to prove in the end fourteen years.[9] Its members soon came round to the view that a scheme on the scale envisaged was well-placed to take advantage of the latest data-retrieval techniques, and this in turn dictated the employment of some full-time staff. A project of such compactness was also likely to hold an appeal for potential funders. By early in 1953 plans had advanced sufficiently to warrant making an approach to the Nuffield Foundation for support, and that December the thrilling news was received that a grant of £10,000 for five years had been awarded. To this sum the Nature Conservancy then undertook to add a further £4,000 extending over the same period, expressly to meet the cost of the specialised equipment that was going to have to be devised, on the understanding that this would pass over to the Conservancy when the Scheme ended.[10] At that time computer mapping, now virtually commonplace, still lay well in the future and it was necessary to have recourse to one of the two leading manufacturers of punched card machinery, Powers-Samas (since absorbed into ICL), to modify the standard tabulating system so that dot-maps could be printed automatically.[11] F. H. Perring, a recent Cambridge graduate student who had already shown interest in harnessing such methods for biological purposes, was recruited as full-time Administrative Officer while S. M. Walters, of the Cambridge Botany School staff, already well-known to many people in the Society, was luckily secured as part-time Director. Accommodation for all the staff of the project and the equipment was in due course provided in the University Botanic Garden.[12]

The Scheme was officially unveiled at the Society's next conference in April 1954, accompanied by a call for volunteers. The response from members, as was only to be expected, was large and immediate; but offers of help soon came in from many other quarters besides. By the end of the Scheme it was calculated that some 1,500 participants had sent in returns (a good many fewer than those who had enrolled, in hopeful enthusiasm); the great bulk of the records, though, were in the event contributed by a *corps d'élite* of some 250. Not all the backsliders were those who had merely lost interest: a few highly expert botanists found the repetitive type of field recording simply not

F. H. Perring, Director (from 1959) of the Distribution Maps Scheme, printing a map on the Powers-Samas machine and holding a pack of individual record cards.

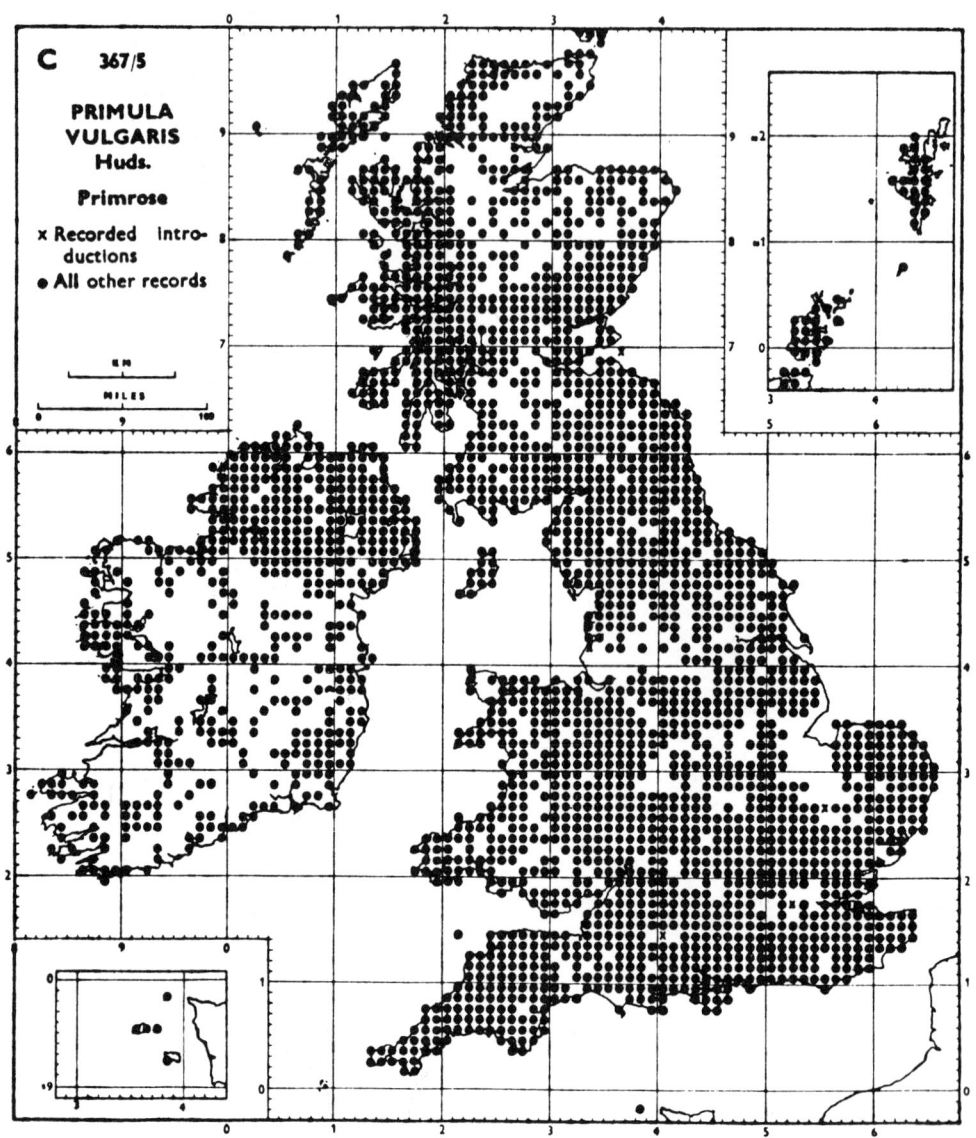

Distribution map showing the 10km squares of the National Grid.

Grid Ref.	LOCALITY	Cliffs N. of Newport Bay Pembrokeshire		WALES	
22/05/1			Date 9/54	V.C. No. 45	
	HABITAT	Slate cliffs and cliff tops	V.C. Pembroke		
			Alt. 10'-60'	Code No. 447	

3	Acer cam	189	Asple mar	365	Carex dio	541	Conop maj	763	Eupat can	976	Hiera pil
5	pse	191	obo	366	distan	544	Convo arv	764	Eupho amy	981	Hippu vul
7	Achil mil	192	rut	367	disticha	548	Cornu san	771	exi	983	
9	pta	193	sep	368	divisa	551	Coron did	772	hel	984	mol
12	Acino arv	194	tri	369	divulsa	552	squ	775	par	988	Honke pep
19	Adoxa mos	195	vir	370	ech	555	Coryd cla	777	peplus	992	Horde mur
20	Aegop pod	204	Aster tri	371	ela	557	Coryl ave	780	por	993	sec
21	Aethu cyn	208	Astra gly	374	ext	562	Coton mic	243		995	Hotto pal
22	Agrim eup	211	Athyr fil	376	flacca	569	Crata mon	783	ang	996	Humul lup
23	odo	212	Atrip gla	377	*flava	572	Crepi cap	784	bor	998	Hydro mor
26	Agrop can	214	has	381	hir	576	pal	785	bre	999	Hydro vul
28	jun	217	lit	382	hos	578	tar	748	con	1000	Hymen tun
32	pun	218	pat	385	lae	579	Crith mar	789	cer	1001	wil
33	rep	216	sab	386	las	586	Crypt cri	796	mic	1002	Hyosc nig
34	Agros git	219	Atrop bel	387	lep	589	Cuscu epith	798	nem	1003	Hyper and
35	Agros can	224	Balde ran	393	nig	592	Cymba mur	799	occ	1006	dub
39	sto	225	Ballo nig	396	otr	596	Cynog off	804	ros	1008	elo
40		229	Barba vul	397	ova	597	Cynos cri	810	Fagus syl	1010	hirsutum
41	Aira car	231		398	pai	603	Cysto fra	813	Festu aru	1011	hum
42	pra	232	Berbe vul	399	pal	607	Dacty glo	816	gig	1013	mon
46	Ajuga rep	234	Berul ere	400	panicea	617	Daphn lau	821	*ovi	1014	per
51	Alche gla	235	Beta mar	401	pauicula	620	Daucu car	823	pra	1015	pul
57	ves	240	Betul pub	404	pen	627	Desch cae	824	*rub	1016	tet
58	*vul	239	ver	405	pil	628	fle	830	Filag ger	1018	Hypoc gla
60	xan	241	Biden cer	407	pse	630	Descu sop	831	min	1020	
62	Alism lan	242	tri	408	pul	434	Desma mar	833	Filip ulm	1023	Ilex aqu
63	pla	243	Black per	412	rem	435	rig	834	vul	1026	Impat gla
64	Allia pet	244	Blech spi	413	rip	640	Digit pur	835	Foeni vul	1030	Inula con
75	Alliu urs	246	Blysm ruf	414	ros	644	Diplo mur	838	Fraga ves	1033	hel
76	vin	247	Borag off	419	ser	645	ten	839	Frang aln	1036	Iris foe
77	Alnus glu	248	Botry lun	421	syl	646	Dipsa ful	841	Fraxi exc	1038	pse
82	Alope pra	250	Brach syl	424	ves	647	pil	845	Fumar bas	1045	Isoet lac
84	myo	251	Brass nap	427	Carli vul	648	Doron par	847	cap	1046	Isole cer
85	pra	252	nig	428	Carpi bet	654	Drose ang	854	off	1047	set
				FOLD	HERE					
87	Altha off	254		431	Carum ver	655	int	862	Galeo lut	1048	Jasio mon
97	Ammop are	256	Briza med	433	Catab aqu	657	rot	867	Galeo spe	1050	Juncu acuti
98	Anaca pyr	262	Bromu com	440	Centa cya	661	Dryop aus	868	*tet	1054	art
99	Anaga arv	269	*mol	444		664	*fil	873	Galiu apa	1057	buf
100	ten	270	mol	446	sca	666	spi	875	cru	1058	*bul
103	Andro pol	273	sec	451		670	Echiu vul	877	ere	1063	con
105	Anemo nem	275	tho	453	pul	673	Eleoc aci	878		1067	eff
109	Angel syl	276	Bryon dio	456	Centu min	674	mul	879	*mol	1069	ger
113	Anisa ste	288	Butom umb	468	Ceras glo	675	pal	840	mol	1070	inf
116	Anten dio	291	Cakil mar	469	sem	678	uni	882	pal	1072	mar
117	Anthe arv	293	Calam epi	462	tet	679	Eleog flu	887	uli	1075	squ
118	cot	296	Calam asc	467		681	Elode can	888	ver	1076	sub
119	nob	298	nep	473	Ceter off	682	Elymu are	891	Genis ang	1080	Junip com
121	Antho odo	2249	Calli agg	474	Chaen min	683	Empet nig	893	tin	455	Kentr rub
123	Anthr neg	303	int	476	Chaer tem	684	*nig	897	Genti *ama	1082	Kickx ela
125	syl	304	obt	477	Chama ang	685	nig	901	*cam	1084	Knaut arv
126	Anthy vul	307	sta	479	Cheir che	687	Endym non	906	Geran col	1087	Koele gra
127	Antir maj	309		480	Cheli maj	689	Epilo adn	907	dis	1098	Lamiu alb
128	oro	310	Calth pal	481	Cheno *alb	692	hir	909	luc	1099	amp
131	Aphan *arv	2248	Calys *sep	484	bon	695	mon	911	mol	1100	hyb
132	arv	311	sep	491	mur	696	obs	914	pra	1103	pur
133	mic	312	sol	493	pol	697	pal	916	pus	1104	Lapsa com
134	Apium gra	313	syl	496	rub	698	par	917	pyr	1107	Lathr squ
135	inu	316	Campa leu	502	Chrys leu	699	ped	918	rob	1108	Lathy aph
137	nod	322	rot	503	par	700	ros	920	san	1112	mon
141	Aquil vul	323	tra	504	seg	705	Epipa hel	923	Geum int	1116	pra
142	Arabi tha	325	Capse bur	505	Chrys alt	708	pal	924	riv	1117	syl
146	Arabi hir	327	Carda ama	506		712	Equis arv	925	urb	1119	Lavat arb
150	Arcti agg	328	fle	509	Cicho int	713	flu	929	Glauc fla	1125	Lemna gib
151	lap	329	hir	511	Circa alp	717	pal	930	Glaux mar	1126	min
152	min	330	imp	513	lut	720	syl	931		1127	pol
153	vul	331	pra	515	Cirsi arv	721	tel	932	Glyce dec	1128	tri
163	Arena lep	333	Carda dra	516	dis	723	var	933	flu	1129	Leont aut
161	*ser	335	Cardu nut	517	eri	726		934	max	1130	his
162	ser	337	nut	520	pal	731	tet	936	pli	1131	
166		339	ten	522	vul	733	Erige acr	940	Gnaph syl	1132	Leonu car
167	Armor rus	341	Carex acu	523	Cladi mar	735	can	941	uli	1133	Lepid cam
169	Arrhe ela	340	acuta	528	Clema vit	740	Eriop ang	948	Gymna con	1137	rud
170	Artem abs	344		530	Clino vul	743	lat	949	Malim por	1139	smi
172	mar	350	bin	532	Cochl ang	744	vag	952	Heder hel	947	Leuco alb
175	vul	355	car	533	dan	745	Erodi *cic	955	Helia cha	1144	Ligus vul
176	Arum mac	357	con	535		748	mar	961	Helic pra	1148	Limon bin
182	Asper cyn	359	cur	537	Coelo vir	753	Eroph *ver	962	pub	1154	vul
183	odo	361	dem	538	Colch aut	758	Eryng mar	968	Herac sph	1155	Limos aqu
185	Asple adi	363	dia	540		762	Euony eur				

Field card, as used in the Distribution Maps Scheme.

A Primrose: from an original water-colour by William Kilburn.
An example of a species whose relative scarcity was highlighted by the Scheme.

to their taste or, in some cases, the print on the record cards too infuriatingly small to cope with. Not for nothing, indeed, was the work quickly – and inevitably – dubbed 'square-bashing': it was essentially a drill, and it required a good deal of self-discipline to perform it accurately and persistently. That so many took to it so readily is proof enough that the work supplied a welcome focus for that high proportion whose botanising had hitherto been without any clear-cut purpose.

In the course of five effective full seasons all but seven of the 3,500 squares in the British Isles were visited, though many of those in Ireland only very slightly and fleetingly. This entailed intruding into many areas which, for one reason or another, had never been explored intensively, at any rate at all recently, and some startling discoveries resulted. Probably the most trumpeted was the finding of a second surviving British station for the almost-extinct Military Orchid (*Orchis militaris*) in quite an unexpected part of the country, Suffolk. Equally unexpected was American Lady's Tresses (*Spiranthes romanzoffiana*) in its first locality in mainland Britain, on the rim of Dartmoor. Had it achieved nothing else, the Maps Scheme would have justified its occurrence by this thorough sweeping of so many corners of the countryside that had hitherto been disdained.

That the Scheme became the focus for some of the Society's field meetings throughout this period goes without saying. For many members this was undoubtedly a deprivation, for the rigorous routine of mapping held few excitements for those who preferred to search just for novelties. The day-long isolation of individual car-loads, each assigned their quota of squares to cover, also made for a loss in sociability. At the same time mapping was not without its own unique adventures. One party, busily crossing off the common species in a square on the borders of Hampshire and Berkshire, was startled to have its cards seized by a suddenly irrupting police patrol. In their absorption the mappers had failed to notice that they were by the perimeter of the nuclear weapons establishment at Aldermaston. Deep was the disappointment of the constables when it turned out that the suspicious-looking script was nothing more incriminating than abbreviated Latin.

At the end of those five years the estimated total of the records that had been accumulated was a million and a half – a thousand for every participant. In addition to the deployment of the great army of card-bearers in the field and checking what they sent in, there had been the hardly less enormous task of ransacking the published literature, national and local herbaria and the card-indexes of county Flora compilers and Recorders in an effort to make the maps of the rarer species as complete and accurate as possible. Even so five years had proved to be barely enough merely for the initial stage of gathering in the data: further funding was now needed to see the Scheme through the data-processing up to the point of producing the finalised maps for publication. Here, thankfully, the Nature Conservancy again stepped in and provided nearly £20,000 more, enabling the work to continue. At that point, in April 1959, Walters bowed out as Director and Perring took over in that capacity in his place.

The publication of the main *Atlas* in 1962 was attended by a final burst of publicity and a celebratory dinner of the Society in London. Printed by Jarrolds and published by Nelsons, it comprised 1,700 maps and 12 transparent overlays. Sales exceeded expectations and the first impression of 3,000 copies was quickly exhausted. These found their way far and wide, even, as a later General Secretary was to discover to her delight, to as unlikely and remote a corner as a bookshop in the Himalayas.

The *Critical Supplement* to the *Atlas* followed six years later. As the title indicates, this consisted of maps – together with, this time, brief individual commentaries – of a miscellany of taxa which had been adjudged to require specially careful treatment, the records utilised being mostly ones that rested on the authority of acknowledged experts.

By that time the Scheme had officially come to its end and both its secretariat and equipment, passing under the Nature Conservancy's wing, had been transferred to the latter's field station at Monks Wood. There they became the nucleus of the Biological Records Centre, stimulating and servicing the forty and more other national mapping schemes that have since been set in train by the BSBI's pioneering example.[13]

In the words that proudly appeared on the *Atlas* dust-jacket, the Scheme was up to that time "undoubtedly the largest survey of its kind ever undertaken in the field of British natural history". It entirely fulfilled the expectations of its conceivers and, in turn, had put the Society itself, indelibly, 'on the map'.

CHAPTER 13

Into the Future

The end of the Distribution Maps Scheme marked the close of a period in the Society's history more generally. As if to underline this, there was a change of General Secretary, a change of President and even a change of Patron – for in 1965, following the death of the Princess Royal, Her Majesty Queen Elizabeth, the Queen Mother, greatly pleased the Society by consenting to take her place. A good field botanist in her own right, the Queen Mother's personal interest in the BSBI has been evident on several occasions since, most memorably in 1972, when those attending a field meeting in Caithness had the honour of being invited to tea at the Castle of Mey and of being afterwards conducted down to the beach to see the Oyster Plant (*Mertensia maritima*) by Her Majesty herself.

The retiring General Secretary was Dony, who had succeeded Lousley seven years before, in 1956, shortly after the Maps Scheme began. With another seven years of service preceding those already, as Field and then Meetings Secretary, Dony's labours on the Society's behalf had rivalled even Lousley's in their duration and their centrality. Kent's, too, were by this time promising to run the records of those other two very close, and in the general change-over of 1965 the opportunity was taken to relieve him of the handling of applications for membership and the upkeep of the address list, a by now impossibly onerous task to combine with his *Proceedings* Editorship. Thenceforward it was gallantly shouldered by Mrs Dony, who was made an additional Assistant Secretary in consequence (to be elevated to a new office of Membership Secretary four years later).

Dony's succession to the General Secretaryship had liberated Lousley for undivided attention to the spiralling demands of the Conservation Committee, of which he continued as the dedicated Secretary. Tragically, though, after only two more years in that capacity, his long and multiply fertile stewardship of the Society came to a sudden and entirely unexpected end.

The incident which caused this was fittingly Watsonian in its surface triteness. A short note had been published in *Proceedings* on the earliest use by botanists of the vasculum. A member whose antipathy to collecting in all its manifestations was particularly extreme used this as a pretext to submit a reply in which he deplored the damage that had resulted from that cause all through the years and urged that it was time now the activity was declared extinct. For Lousley (and at least one other Conservation Committee member) this was just the kind of sweeping exaggeration on this – to him – highly sensitive subject that he had long been at pains to counter, and he took the view that for the Society to allow it to appear in print would be seen as an endorsement of it officially. The Editor of *Proceedings*, on the contrary, considered that to refuse publication would be tantamount to censorship; he therefore felt bound to permit the note to appear while at the same time making it quite apparent that it was merely one member's personal view. Lousley insisted that even to do this would make his position impossible, for the note could be seen as implying that for an active collector to serve on a committee devoted to conservation showed a lack of principle. The Council was asked to adjudicate; but on finding no support there for his stand, Lousley thereupon resigned.

Not long after that, very fortunately, the newly-established Council for Nature, which was brought into being primarily as a means of enabling the many national and local societies to speak with a single voice in the matter of conservation, found itself in need of an honorary secretary; and Lousley, with his many years of experience in that direction and his proven administrative flair, made for such a body the ideal catch. Nevertheless he was not to escape the toils of the BSBI all that easily. For in less than three years, in 1961, the Society was welcoming him once again back into office, only this time as its President: the first amateur to be so honoured in what was deliberately to be henceforward an amateur/professional alternation.[1]

The resignation of Lousley had coincided in turn with a change of Treasurer, J. C. Gardiner taking over from Swann. One eventual result of this was that the Society's finances came to be administered for several years from the appropriately-named Thrift House, where – shades of Bailey and Ralli Brothers – the BSBI became the pleasingly incongruous bedfellow of a business combine which embraced the British Shoe Corporation and Selfridges.

The inflow of royalties from the sales of the *Atlas* led to a sharp improvement in the Society's financial position from 1963 onwards. Three or four years later an unintended falling-behind in the appearances of *Watsonia* made the position temporarily even better, and in the meantime substantial amounts of interest were earned on the unexpected funds. This untypical flushness allowed the Society to indulge itself at last in a modicum of paid assistance, sufficient to relieve the officers of the drearier portions of their

HM Queen Elizabeth, The Queen Mother returning with a party of members from a visit to *Mertensia maritima* on the shore at the Castle of Mey, Caithness, 1972.

Oyster Plant (*Mertensia maritima*). Engraving by Dillenius.
This species is withdrawing northwards and is now confined to the coasts of northern and western Scotland and north-east Ireland.

work. Buying into common services shared with other, similar organisations was the obvious solution here, a convenient half-way house which made it possible to sidestep the invidiousness of a situation where paid staff are intermixed with volunteers. In the years since, more and more of the administrative routine has been offloaded in this manner, so that still today, with a membership approaching 3,000, the BSBI continues to put many much smaller societies to shame by operating internally on the voluntary principle alone.

A further benefit from the Maps Scheme was a marked increase in the Society's self-confidence. It had established that it could count on the active support of a large proportion of the membership; it had learned a great deal about publicity; it had also grown impressively larger, having almost doubled again in size (to nearly 1,600) during the ten-year period.[2] It was in a mood for some testing of this new-found strength.

Within a year of the close-down of the Scheme a worthy challenge to this muscle obligingly appeared. The Tees Valley and Cleveland Water Board, it was learned, was planning to construct a large impounding reservoir at Cow Green in Upper Teesdale, which seemed to impose a devastating threat to the remarkable relic flora for which that valley has long been so renowned. In February 1965 the Society in conjunction with Northumberland and Durham Naturalists' Trust set up a Teesdale Defence Committee in response. Gardiner became its Treasurer and gave it unsparingly of his energies and masterminded its tactics. But it was to prove a far lengthier and costlier battle than anyone at the outset supposed, for instead of becoming the subject of a public enquiry as usual, the proposal was introduced as a Private Bill in Parliament. Fighting this through the successive procedural stages entailed ferociously heavy legal and other costs, but fortunately for the Society these were entirely covered by donations, very many of them from members, in response to a series of public appeals which brought in £23,000 in all. In the end the Bill went through and to that extent the battle was lost; but important concessions had been obtained, a very large sum was made available for conservation research by ICI, the chief industrial beneficiary, and the weight of publicity aroused had been such that promoters of any similar projects in future were likely to embark on them more warily. History, moreover, had been made: the countries' botanists had shown the wider world for the first time that they were prepared to fight, and fight extremely hard, to protect what they valued. In their turn botanists discovered, not without a shock, that that wider world would not necessarily accept without challenge assumptions about rarities and the threats presented to them that in botanical circles had long had the standing of commonplaces.[3]

In championing the cause of Teesdale, the Society may have appeared to some to have found itself an enduring new role, a much-needed fresh focus for its energies, that could serve to fill the void that had been left by the completion of the Maps Scheme. To assume such a role, however, it would clearly require far greater resources than the subscription income a society of its kind and therefore size could credibly count on having. It would need in fact to become a very different kind of body altogether.

There were, as it happened, voices just around then that were urging upon the Society an all-out drive for expansion. Heartened by the momentum brought about by the Maps Scheme, impressed by the leap since the 1950s in the size of the following for natural history in general, convinced that field botany was forfeiting its rightful share of this by a failure to present itself attractively, a group of leading members came together to plan a new type of publication by means of which in large part they saw the Society greatly adding to its numbers. This was to be a well-produced illustrated magazine, with the title of 'The Botanist'. Published by the Society and issued quarterly or even monthly at a modest three and sixpence, it would indeed have been a most appealing purchase. As the President wrote in a specimen foreword,

> Interest in plants extends far beyond those who claim to be botanists and this new publication will satisfy a need for which there is at present no provision. It will also be appreciated by readers who are already experts in a narrow field of botany, who will welcome the opportunity of obtaining up-to-date information about other branches in accounts free of technical jargon.

The range of items it was planned to cover included reports on work in progress, accounts of conservation initiatives and guidance with naming plants in difficult groups. "Young people will make their own contributions, and there will be stories about botanists, and about the places where plants grow." In short, this was to have been "a lively, up-to-date magazine, easy to read, and covering all aspects of our native flora".

Alas, it never got beyond the 'mock-up' stage. A costing quickly showed that it had no hope of proving economic. At the low cover price that was considered essential it would have required a continuing subsidy from the Society that was far beyond its means.

By the time that fact was realised opinion within the Society had in any case hardened against the proposed dash for growth. Out of a complex and protracted series of discussions running from the autumn of 1963 for a full three years the consensus finally emerged that the Society would be wrong to embark on a course that imperilled its existing character. A rapid increase in membership seemed likely to be obtained only by lowering standards to a

Design for the cover of the abortive periodical 'The Botanist'.

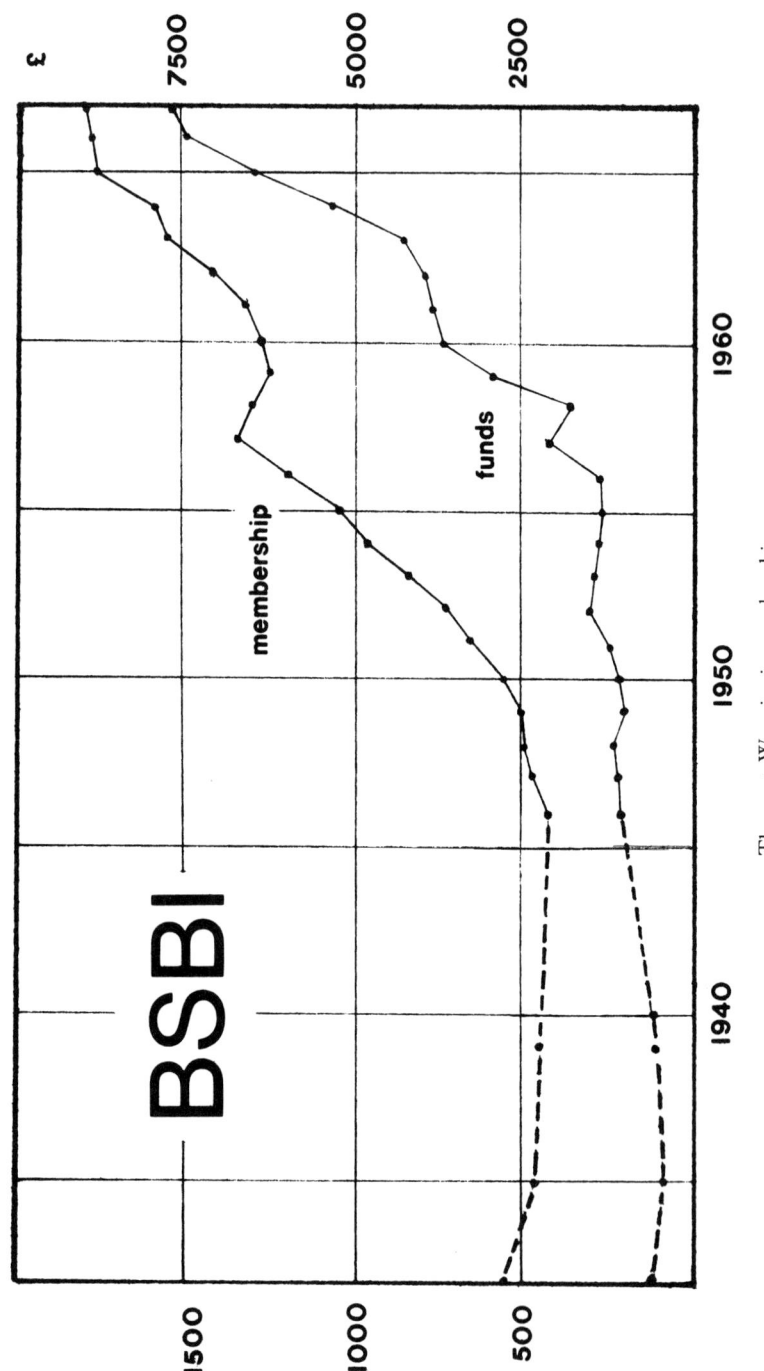

The post-War rise in membership.

J. E. Lousley, the architect of the dramatic rise in membership, photographing alien plants in Hampshire.

level that was unacceptable. While it was true that the fight over Teesdale had demonstrated the need for as powerful a voice as possible if the British flora was to be conserved effectively, that particular battle, it was generally acknowledged, had been exceptional in every way. The BSBI must remain a scientific society first and foremost, the prevailing view insisted: of high importance though it was, conservation could never be more than one of its concerns and ought to be firmly subordinate to the overriding aim of the advancement of our understanding of plant distribution and systematics.[4]

At the same time it was also generally recognised that the Maps Scheme had left behind a vacuum as far as fieldwork was concerned that was urgently in need of being filled. An outburst of new county Flora projects conceived on the same principle (but mostly with the much smaller 2 km square, or 'tetrad', as their mapping unit) was to some extent absorbing the enthusiasm that had been released; but these were private initiatives and as such hardly absolved the Society from finding some substitute activity for its now scheme-hungry membership. The obvious solution seemed to be to follow the example of the students of birds and to launch a continuing programme of co-operative enquiries.[5] In the wake of the Maps Scheme, field botany had every appearnce of having reached that stage at which field ornithology had taken off at the start of the 1930s so spectacularly. If it could succeed in imparting a similar vigour by a well-directed series of Maps Schemes in miniature, the Society might yet achieve substantial growth without having to depart from the straight and narrow of research. Unfortunately for this thesis, there could not be the same objective for botany as there had been for ornithology, of bringing into being a permanent research institute of its own; for having conceded to the Nature Conservancy the signing-up of the home team, the BSBI had kicked its own ball into what amounted to perpetual touch. For the foreseeable future it was unlikely that it would be able to rise to a replacement for this from within its own resources.

In retrospect, it can be seen that the identification of mapping with those types of fieldwork that form the staple of the 'network research' of ornithologists was made too optimistically. Mappers are not required to take measurements or to form estimates of numbers. The mere listing of species that occur in particular tracts of country is a sharply simpler task than such scarcely demanding alternatives. In his efforts at vice-county recording Watson had no difficulty in enrolling helpers; but when in 1850 Henfrey had tried to interest the members of the BSL in testing the validity of named entities by cultivating them in as many different parts of the country as possible, there appears to have been no response. To go beyond listing entails the leaping of a by no means negligible mental barrier. It is scarcely a matter for surprise, therefore, that the programme of special enquiries set in train in 1968 and continued down to the present has seldom appealed to more than a small proportion of the membership or that the exceptions have been those, such as the Mistletoe Survey of 1970–71, where the topics have called for little more than straightforward mapping. As a line of work it has been valuable and productive; but there is no escaping the sad truth that, despite the popularity of one or two individual projects, it has substituted for the Maps Scheme in only a minor fashion.

It is conceivable, though unlikely, that more might have come of this initiative had the Society been enlarged on the scale that some had been proposing. Even if the increase in the number of participants was not nearly in proportion, enough extra ones might have been yielded by an influx of such dimensions to give rise to the necessary 'critical mass'. The huge expansion in numbers that took place in ornithology, after all, had been accompanied by an escalation in research achievement.

However, it was not for the benefits that it could bring to research that the super-enlargement had been advocated. And even had it been, it must be doubted if the resistance would have been much less. For at the root of that resistance lay the fear that the Society would be irreparably damaged by losing as a result most if not all of the professional botanists among its members. The increasingly happy and fruitful co-operation between professionals and amateurs which had characterised the Society since the early 1930s was now hailed on all sides as its single most precious feature. Already, the part that professionals played in its affairs was noticeably less marked than ten years earlier. Any lowering of the collective sights might only hasten the departure of those who remained.

It was all the more ironical, therefore, that shortly after the importance of this amateur–professional partnership had been asserted so emphatically the divide between the two was worryingly broadened by the abandonment of *Proceedings*. The amateurs had come to regard this as the journal which reflected their interests especially, preferring it to the extent that one contributor even went so far as to veto for *Watsonia* material he had submitted which was suitable for either.[6] In 1967, however, a change to new printers had made its production a good deal more costly and this occurred just at the time when Abstracts from Literature, its most valued feature, had begun to proliferate as a result of more ambitious coverage. Within two years this was placing the Society's finances under insupportable pressure. Consequently, by a majority vote on the Council, which split very largely along amateur/professional lines, the decision was taken to merge both journals in a somewhat widened *Watsonia* and to hive off the Abstracts, in a remodelled form, in a new periodical of their own to be published annually. Thus, very abruptly, the literature issued to members took on a much more technical slant. For naturally the editors of the new combined journal, who were preponderantly the editors of the previous *Watsonia*, were averse to any reducing of a scientific standard that had been built up so diligently over so many years; yet as long as this was to be so dauntingly high, many fewer amateurs than previously would venture to submit their work for publication by the Society. The existence of just the one, forbiddingly learned, main periodical hardly served as an inducement, moreover, to that less learned majority who needed to be recruited as members. Three years later, therefore, in 1972, *BSBI News* was started to help redress the balance. Though a newsletter, however frequent and however rich in its contents, is no real substitute for a full-scale journal, this periodical has turned out to be extremely popular. In particular its more informal style has proved especially conducive to the reporting and describing of alien casuals, study of which had previously been inhibited by their inherently weaker claim than the more established flora to costly journal space; in addition, it has afforded a needed outlet for the airing of controversial views by the publication of letters. All the same it cannot be pretended that the mix of publications is now as satisfactory as it was previously. The sacrificing of *Proceedings* represented an undoubted retreat from a high point that economics may never allow to be regained.

Remarkably, since the immediate post-War years the amateur/professional ratio has hardly shifted at all. At present around 375, or 15 per cent of the personal members, identifiably follow botany as their profession (or used to before they retired). Almost certainly that is not the full total, for some more must lie concealed in the membership list under private addresses – in particular biology teachers in schools. But that problem of classification also existed back in 1950, and then the percentage was almost exactly the same. Despite the enormous expansion in higher education which has occurred in the intervening years the Society has not become a less preponderantly amateur body in any significant degree. As fast as the teaching posts in the universities and colleges have multiplied, as fast as new avenues of employment like the Nature Conservancy and the field centres have appeared, the following for field botany among the public at large and its representation as a result in the BSBI's ranks have expanded in step.

The distinction between amateurs and professionals is not, though, as clear-cut, at any rate in a society such as this, as the conventional dichotomy accustoms one to think. In reality there is a continuous spectrum, thanks to various intermediate groups which moderate the contrast between the two extremes. Many of those who work in the major taxonomic institutions, for example, have traditionally been as close, if not closer, to the amateur world as they have been to the world of the universities. For the best part of a century taxonomy was so out of favour academically that there were almost no fellow spirits in the latter sphere at all. Those who did pursue it professionally mostly owed that fact to an appetite for British field botany well predating their student days – an appetite which many of them were more than happy to keep up. Reinforcing that tendency was the shortfall in the production of graduates in botany for the taxonomic posts in government service, which had the effect of forcing a temporary crumbling of the normal certification barrier to the extent that in the years on either side of the Second World War a minor recurrence of the old-style recruitment of amateurs took place. More recently the ranks of the amateurs have been swelled in turn by those pursuing careers in other walks of science who have taken up field botany as a recreation. There are more amateurs around in consequence who are acquainted with biometrics and who have at least seen a chromosome.

It is this degree of interpenetration between the amateur and the professional outlooks that constitutes the special strength of the present-day BSBI (and indeed other national botanical societies in Britain that are concerned with other groups of plants such as ferns, mosses and lichens), rather than the presence of a sizeable professional element among the membership. Indeed there is nothing particularly remarkable in the fact that amateurs and professionals should have a corporate coexistence. Nor is it particularly remarkable either that they should contrive to work harmoniously together. Fruitful collaboration between the two continues to be the proud boast, at any rate in Britain, of the observational sciences in general. In astronomy, in meteorology, in archaeology, in all the old 'natural history' studies, there is still a respected role here for the amateur, in a way that there is no longer, alas, in most other areas of scholarship, and in almost all other countries in the world. In the Geologists' Association there has even been reproduced, by a strikingly parallel process of development, a body which closely resembles the BSBI in all its general functions and features, including the selfsame amateur–professional symbiosis. But it must be doubted whether in any other field accidents of history have enabled the splicing together to be carried so far so enduringly and so successfully.

The number of professionals the Society has chanced to attract *in proportion* to the number of amateurs may well be a subsidiary ingredient in that success. Too heavy a weight of professionals not only makes the amateurs feel smothered and saps their will to emulation, but it also imports an atmosphere of career concerns and the extraneous tensions that arrive with these. Had the

BSL survived into the early years of the present century, it would surely have shared the fate of its entomological sister, which was so overwhelmingly invaded by the jarring interests and language of the newly-arisen exponents of the applied aspects of the science that the amateurs gradually slunk away and inflated the South London Society into an alternative national body. Had F. H. Ward only put into effect his 1890 notion of a merger of the three clubs that jointly encompassed the study of the taxonomy and distribution of the countries' vascular plants, a twin might have been produced of that other offspring of a merger in those years, the Botanical Society of America. Just as that society, as a result, came to cover all aspects of the science, so a British Isles counterpart might have done so too; so, equally, might it have matured into the body representing the emerging profession. Where today, one wonders, would the main mass of amateurs find their home had events taken such a course?

A society composed of amateurs more or less exclusively would never have risen to the heights of a journal of the quality of *Watsonia*. Equally a society made up for the most part of professionals would never have developed such an extensive and varied field meeting programme or such serried ranks of human machinery for monitoring distributional change. For perfect health a goodly number from both categories is needed. For both have their particular strengths, their complementary skills to contribute on one side or the other of the unspoken division of labour: the professionals shouldering the bulk of the editing of the scientific journals and the directing of research, the amateurs concentrating on the compiling of records and taking the greater share of the administrative routine. In this way the feeling of a joint endeavour is subtly enhanced.

To work in harmony with amateurs requires of the professional a particular kind of patience. He needs to bear in mind that matters which interest him intensely will not necessarily interest them in the slightest, that what they choose to study may be but tenuously of profit to science. He has to learn to live with the fact that the archetypal amateur likes to amass data for the pleasure of the amassing, regardless of theories and concepts which might bring to his efforts a more beneficial direction. For amateurs are by no means necessarily scientists *manqués*; few of them are concerned even to carry out research in the sense that the professional understands that term. For most of them their work consists in just an orderly observing and is an end unto itself. If it should happen to prove of use to the professional for his purposes, well and good; but no one should be surprised if amateurs persist in seemingly pointless courses from which there can be no deflecting them.

Yet the amateur element in societies like the BSBI has more to offer the professional than the often random turning up of helpful raw data, more even than that higher proficiency whose products stand comparison with, and valuably supplement, those of his own research; for he can acquire such assistance irrespective of whether there is a society that exists to promote it or not. In so far as he finds the need to operate within the frame of an organised body, however, he is beholden to the amateurs for a further, very special contribution. This is for the extra depth of commitment that so many of them tend to bring to their membership. For, unlike the majority of professionals, they see less point in spreading their allegiance among a number of national societies and are more inclined to identify with just the one in particular instead. For them, their membership is not in competition with those many other means that are open to professionals of keeping in touch with those who share their interests. For them, far more than for professionals, a national learned society constitutes an oasis, a source of refreshment that gains in attraction by the extent to which it stands on its own. The consequent clustering around it of so much undivided loyalty makes for an exceptionally strong morale, which makes in turn for an impregnable stability. Herein lies another of the BSBI's hidden strengths.

Nevertheless, in spite of the advantages to the Society, and to themselves, that this leavening of professionals brings, it would be wrong to assume too airily that just because it has lasted so long it will last in at least the same measure indefinitely. The taxonomic tide that swept in during the 1950s has for some years now been receding: there are no grounds for confidence that those who were borne into the Society on its current will be replaced as they begin to fall away. In so far as a fresh generation may come after, there cannot be much confidence, either, that the interests of those newcomers will similarly be in the vascular plants of the British Isles. The post-war surge was such that marvellously few of the inveterate awkward corners escaped its pervasive wash. One by one the difficult groups and the controversial species have come in for a conclusive cleaning-up, with the result that little now seems to remain that can offer enough by way of a challenge. The time may have arrived when the biosystematists will decide to withdraw and direct their attentions elsewhere.

Alternatively, of course, the Society itself may opt for a broadening in its scope. It is, after all, only an accident of history that it has taken the British Isles as the limit of its coverage geographically. Indeed, that it runs even as widely as that could be said to be an accident too. Watson was content to confine his efforts to Great Britain alone, and there was a self-sufficiency inherent in Irish field botany that was to be deeply consolidated by Praeger.[7] As late as 1910 Irish members of the BEC totalled precisely two. Druce had little enough grounds, therefore, for styling the grander body of his ambitions 'of the British Isles'; but by that almost casual action he effectively pre-empted the emergence of a separate Botanical Society of Ireland. Not that botanists there were ever anything like as numerous in proportion as in the sister island. Even in recent years, despite the introduction of biology into the curricula of the Republic's schools and despite the liveliness of the Society's Irish Committee, Ireland's share of the membership has yet to rise above about three per cent, less than half the share that Scotland has been able to achieve. This ratio between the two has been more or less constant since the War.

Since the 1950s and the launching in those years of the great *Flora Europaea* project attempts have been made to shift the Society out of this parochialism. The Presidency of S. M. Walters in 1973–5 was particularly noteworthy in that connection, the wider European context of the Society's traditional concerns being chosen as the topic of a major conference at his instance. A little earlier, in 1968, the BSBI had been drawn into discussing with several of its Continental counterparts a proposal for a federation of Europe's botanical societies, a proposal which, though it eventually came to naught, was taken sufficiently seriously for the General Secretary to accept the invitation of the Società Botanica Italiana to attend its annual conference. Yet although botanical holidays abroad have increasingly come into vogue and field meetings on the Continent feature more and more in the annual programme, there has so far been no move to extend the Society's boundaries officially. The flora of the British Isles thus continues to remain the almost exclusive focus.

Another possibility would be for the Society to cease to confine itself to taxonomy and floristics – or to vascular plants alone. Yet that seems a much fainter prospect, for catering for the other aspects of botany are other, well-established bodies which leave no room for a late-coming competitor and which it is hard to believe would seriously entertain proposals for amalgamation.

That is not the only way, however, in which the Society could alter its very structure: it could also revert to its origins and assume a metropolitan character all afresh. Once again, it was only an accident of history that left it without premises and collections, without even a programme of regular evening meetings. On the one hand that has been a blessing, sparing it the crippling financial burdens that more anchored societies have been beset with, but on the other hand it has deprived that forty per cent of the membership which lives in the south-east of England of an off-season forum that it might well have put to valuable use. There remains a gap here which might one day be filled.

Military Orchid (*Orchis militaris*): from an original water-colour by E. J. Bedford now in the British Museum (Natural History). The discovery of this orchid in a second British locality was a surprise by-product of the Distribution Maps Scheme.

The Society need not necessarily adopt any of these courses: it may decide that it prefers to stay as it is. The winds of fate, though, may determine otherwise and force unwanted change upon it. Yet however hard it is buffeted, one thing is certain: it will surely endure. It has come a very long way in the last fifty years, let alone in the full century and a half of its existence. Any learned society that has managed to survive through a period so lengthy must surely have acquired a staying-power that will continue to stand it in good stead. It fills a niche – a niche which one cannot suppose will disappear. While it may be too much to hope that field botany will ever again enjoy that share of the public attention that came its way in the nineteenth century, it is hard to believe that its appeal will not be sustained. As leisure continues to expand, as higher education spreads still further, the following for pursuits such as this seems bound to go on growing. The BSBI may very well change and may very well acquire a different focus or a quite other pattern of activities; but it cannot help but thrive, and it will surely continue to do so for another hundred and fifty years at least.

NOTES

Abbreviations used:

I MANUSCRIPTS

BAB	Babington correspondence, Botany School, Cambridge
BAK	J. G. Baker correspondence, BM-B
BAL	Balfour correspondence, Royal Botanic Garden, Edinburgh
BM-B	British Museum (Natural History), Dept. of Botany
BM-G	British Museum (Natural History), General Library
BM-Z	British Museum (Natural History), Dept. of Zoology
BSBI	The Society's own archives, BM-B
DR*	G. C. Druce papers, Botany School, Oxford (*Followed by the relevant box-file number, pending the sorting and cataloguing of this very extensive collection)
Flower	Reminiscences of T. B. Flower noted down in August 1882 by W. Bowles Barrett, Weymouth Central Library (BB 581 94233/BA1)
HOO	Hooker correspondence (English letters), Royal Botanical Gardens, Kew
LIV	T. B. Hall papers, Merseyside County Museums, Liverpool
PE	W. H. Pearsall correspondence, BSBI
THO	H. Stuart Thompson papers, Bristol University Library
WIL	W. Wilson correspondence, Cryptogamic Herbarium, BM-B

II PRINTED WORKS

Babington	*Memorials, Journal and Botanical Correspondence of Charles Cardale Babington*, ed. A.M.B[abington]. Cambridge, 1897
Gage	A. T. Gage, *A History of the Linnean Society of London*. London, 1938
Gunther	A. E. Gunther, 'The miscellaneous autobiographical manuscripts of John Edward Gray (1800–1815)', *Bull. Brit. Mus. (Nat. Hist.), Hist. Ser.*, 6 (1980) 199–244
Proceedings	*Proceedings of the Botanical Society of the British Isles*
Report	*Report of the Botanical Exchange Club* (under its successive variations in title)

From Chapter 1, The Origins

1 The first, and fullest, account of this Club is to be found in George Pasti's 1950 University of Illinois doctoral thesis, 'Consul Sherard: amateur botanist and patron of learning, 1659–1728'.

2 D. E. Allen, 'John Martyn's botanical society: a biographical analysis of the membership', *Proceedings*, 6 (1967) 305–324. By a remarkable coincidence one of the twenty-three members, a Walter Tulliedeph, was the ancestor of a member of the BSBI at the time that paper appeared: Mrs A. H. Gurney, of Walsingham Abbey, Norfolk (see my note in *Watsonia*, 15 (1984) 162). Some supplementary information has since been published in *Soc. Hist. Nat. Hist. Newsletter*, no. 24 (Feb. 1985) 9–10.

3 The Linnean's equally generalist predecessor, the Society for Promoting Natural History, founded six years before, in 1782, early became dominated by geology.

4 Its informal offshoot, the Linnean Club, started to hold excursions some ten years after its founding, in 1811 (T. M. Harris, 'The minute books of the Linnean Club', *Biol. J. Linn. Soc.*, 3 (1971) 343–368); but membership of that was even more exclusive than of the Society itself.

5 J. E. Gray, 'Sowerby's "English Botany"', *J. Bot.*, 10 (1872) 374–375.

6 Ironically, Smith at first thought S. F. Gray's name was a pseudonym of his arch-enemy R. A. Salisbury, who had already suffered a hounding from the Linneans for having dared to advocate the Natural Method in print. Salisbury is known to have helped the Grays by lending them manuscripts for use in preparing their book, so he was not indeed innocent altogether (D. J. Mabberley, 'Generic names published in Salisbury's reviews of Robert Brown's works', *Taxon*, 29 (1980) 597–606).

7 The more particular pretext, almost unbelievable in its pettiness, was the claim that Gray had deliberately insulted the President by a reference to *English Botany* that implied that it was the work of Sowerby – its illustrator – alone (A. Macleay to Smith, May 4th, 1822 – cited by Gage, p. 29; see also Gray, note 5). Smith himself was horrified by the blackballing, which was to prove unique in the Linnean Society's history.

8 For a fuller account see my paper, 'The naturalist tradition in the universities of Britain', *University Botanic Garden, Cambridge, 150th Anniversary Celebrations* (Cambridge, 1981), pp. 7–18. I have also set out the train of events in *The Naturalist in Britain: a Social History* (London, 1976), pp. 106–109.

9 Doubtless the J. 'F'. Cooper listed as one of the inaugural Vice-Presidents of the London Chemical Society, an ephemeral body suggestively similar to the BSL in its aims and social composition (see W. H. Brock, 'The London Chemical Society 1824', *Ambix*, 14 (1967) 133–139). Later, he was to serve as Treasurer of the more

solid and enduring Chemical Society of London, founded in 1841 by his pupil Robert Warington. Fuller details of his career may be found in *Quart. J. Chem. Soc.*, 8 (1856) 109–110 and in R. Bud & G. K. Roberts, *Science versus Practice: Chemistry in Victorian Britain* (Manchester, 1984), pp. 27–30.

10 *Cybele Britannica* (Thames Ditton, 1847), Vol. 1, p. 279: "that mass of confusion and blunders – the Flora Metropolitana".

11 The only substantial letters of Cooper's that have been traced are two written to the editor of the *Magazine of Botany and Zoology* in December 1836 and February 1837, now among the Jardine papers in the Royal Scottish Museum. Regrettably neither refers to the Society.

12 G. W. Francis to W. Wilson, April 29th, 1935: Wilson correspondence, Warrington Central Library.

13 For fuller details see W. J. Hooker, 'Some account of a society lately established in Germany', *Edinb. J. Science*, 7 (1827) 23–29. Sophie C. Ducker ('History of Australian phycology: early German collectors and botanists', in *History in the Service of Systematics*, ed. A. Wheeler & J. H. Price (London, 1981), pp. 43–51), has recently pointed out that in Germany at this period there was none of the private and official patronage of overseas collecting that the British and the French could rely on. The *Unio Itineraria* thus arose in response to a special national need.

14 D. E. Allen, 'H. C. Watson and the origin of exchange clubs', *Proceedings*, 6 (1965) 110–112.

15 *Mag. Nat. Hist.*, 4 (1831) 166.

16 J. H. Balfour, 'Sketch of the life of the late Professor Edward Forbes', *Trans. Bot. Soc. Edinb.*, 5 (1858) 23–41.

17 See my note in Newsletter No. 4 (Nov. 1979) of the Society for the Bibliography of Natural History, p. 5. See also G. Wilson and A. Geikie, *Memoir of Edward Forbes, F.R.S.* (Edinburgh, Cambridge & London, 1861), Chapter VII.

18 Watson to Sir William Hooker, May 3rd, 1844: HOO.

19 Watson to Babington, March 27th, 1844: BAB.

20 Like the Linnean, the Edinburgh society also acquired a greater social *cachet*. For example, the cultivated Lord Braybrooke (under whose aegis Pepys' diary first saw the light of day) put his daughter into the BSE, while the BSL rceived his head gardener.

21 Of the 371 traced as putative members of the London society in the course of its existence, only 66 (18 per cent) were ever members of the Edinburgh one – and not necessarily simultaneously.

From Chapter 2, A Society is Born

1 Intriguingly, of all the periodicals reporting the Society's meetings only the *Gentleman's Magazine* chose regularly to mention the dinners (doubtless because of its greater concern with the chronicling of the social side of things). This is jolting evidence of how selective the various sources are in their accounts.

2 Like the Duke, White was also a leading freemason: Grand Secretary for forty-seven years, and eventually Grand Sovereign, of the Masonic Order of Knighthood (obituary in *Freemasons Mag.*, April 14th, 1866). But despite the claim of Bernard Faij ('Learned societies in Europe and America in the eighteenth century', *Amer. Hist. Rev.*, 37 (1931) 255–266) that "almost all of the learned societies which flourished in the second half of the eighteenth century were imbued with the spirit of Masonry and often worked in close cooperation with the local lodges", his connection with the Society is more likely to have been innocently scientific than the product of a hidden recruiting network.

3 Chatterley was about Cooper's age and may have been a friend from schooldays. Many records of his feature in Cooper's *Flora Metropolitana*. The Society's inaugural Prospectus (reproduced in *Mag. Zoo. Bot.*, 3 (1836) 501) lists Cooper's home as his address too, so maybe at that time he lodged there. Significantly, both he and Cooper's father were then teaching chemistry at the City of London Institute in Aldersgate Street. Later, when around thirty, he qualified in medicine and after various posts in London went out to Australia, dying there soon after arriving, at only forty-one.

4 A letter from George White to T. B. Hall (September 6th, 1836: LIV) implies that a public meeting had been advertised for that next day. Presumably it was postponed.

5 *The Times*, October 14th, 1836, p. 3, col. 3 (reprinted in *Gardener's Mag.*, 12 (1836) 693–696). *The Times* carried reports of the first four meetings, but after that it evidently decided that the new society was not of a sufficient stature to deserve notice in its columns. The *Morning Chronicle* was less speedily disdainful and in later years proved willing to publish from time to time material it adjudged of general interest. Otherwise, meeting reports seem to have been confined to the more or less learned periodicals. An impressive number of these, eleven in all, were involved at one time or another but only four on any prolonged and regular basis. These were the *Gardeners' Chronicle*, the *Phytologist*, the *Athenaeum* and the *Literary Gazette*. Most of these apparently relied on the Secretary for their reports (as was customary with other societies), but those in the *Literary Gazette* were contributed independently and include lengthy abstracts of papers not to be found in the others. The *Gentleman's Magazine* drew on this latter source too, but for the most part reported the Anniversary Meetings only. By far the most informed and discriminating notices were those of the *Magazine of Zoology and Botany* and its successor, the *Annals and Magazine of Natural History*; these, however, were very selective and spasmodic and were discontinued entirely after 1844. They, too, include some lengthy abstracts which appear nowhere else.

6 A copy of the rules in their eventual printed form (1837) survives in the library of BM-B.

7 For a detailed account see my paper, 'The women members of the Botanical Society of London, 1836–1856', *Brit. J. Hist. Science*, 13 (1980) 240–254. It was perhaps due to its example that the short-lived Ornithological Society of London, founded the next year, also admitted women as members (at half the male subscription rate) and welcomed them at its monthly meetings.

8 A. Heathcot, *Mag. Zoo. Bot.*, 1 (1836) 502.

9 Later altered to Fridays. After 1844 the meetings were cut down to only one a month throughout the year (with a gap around Christmas).

10 Heathcot, l.c.

11 Of the 38 subscribing members enrolled by the time of the meeting following, no fewer than 10 dropped out within the next twelve months – a surely excessive number?

12 However, he had fallen out with the Zoological Society and its authoritarian chief officer, N. A. Vigors (Gray to W. Swainson, August 5th, 1830: Swainson correspondence, Linnean Society of London). This added to the isolation already inflicted on him by his exclusion from the Linnean Society.

13 D. C. Macreight. His *Manual of British Botany* was published in 1837.

14 In this he was being over-complacent. The Entomological Society acquired 117 members in its first eighteen months, a total which the Botanical Society had to wait three more years to reach.

15 Departmental reports, Vol. 45 (1835–45), unnumbered section: BM-Z.

16 One unexpected absentee was Cooper's mentor, J. Forbes Young. He must surely have been approached but declined to stand.

17 The meteorological connection has been overlooked because of the misprinting of 'Met.' as 'Med.' in the Society's 1839 *Proceedings*. For further details see my note in *Watsonia*, 12 (1979) 390.

18 G. J. Symons, 'The history of English meteorological societies, 1823 to 1880', *Quart. J. Met. Soc.*, n.s. 7 (1881) 65–98.

19 F. Bossey to Hooker, undated [early 1837?]: HOO.

20 H. C. Watson, 'On the credit-worthiness of the labels distributed from the Botanical Society of London', *Phytologist*, 2 (1847) 1000–1015.

21 *Proc. Bot. Soc. Lond.*, 1 (1839) 57.

22 Watson to Babington, March 27th, 1844: BAB.

23 The register of baptisms of Bath Abbey confirms the suspicion that he was the son of the William Simmonds Chatterley who features in the *Dictionary of National Biography*. According to the latter his mother, in turn, was a well-known comedienne.

24 Public Record Office, Court of Bankruptcy papers B.3/3537.

25 *The Times*, October 14th, 1836, p. 6, col. 3.

26 *Proc. Bot. Soc. Lond.*, 1 (1839) 94.

27 H. B. Fielding originally planned to bequeath his very rich and extensive collection to the Regent's Park society (Fielding to Hooker, February 9th and March 18th, 1842: HOO), but because of its inability to ensure the stipulated accommodation he became the principal creator of the Oxford University Herbarium instead. There is some evidence that the RBS herbarium was one of those to benefit from the auction of the Botanical Society's collections in 1857. It was eventually donated to the British Museum (Natural History) in 1933.

28 For a definitive account see G. Meynell, 'The Royal Botanic Society's Garden, Regent's Park', *London Journal*, 6 (1980) 135–146. The two in turn should be distinguished from the Medico-Botanical Society of London, founded in 1821 by the egregious John Frost (see B. Hill, 'A Georgian careerist', *The Practitioner*, 188 (1962) 262–266). This had a much grander membership than the BSL, with scarcely any overlap.

29 In fairness it should be added that the Edinburgh society's exchange activities also suffered from serious shortcomings. Babington (to J. F. Duthie, October 8th, 1874: Babington, p. 372) recommended that most of its early duplicates be destroyed because of the untrustworthiness of the labels.

30 See my note, 'The first woman pteridologist', *Br. Pterid. Soc. Bull.*, 1 (1978) 247–249.

31 Spreading lengthy papers over more than one meeting was, however, a not unfamiliar device in learned societies at this period. An extreme case was a Commentary on the *Hortus Malabaricus* by Francis Buchanan-Hamilton, which the Linnean Society began to listen to in May 1821 and only finished devouring thirty-two years later. On fourteen occasions it was the sole fare provided (Gage, p. 114). Such papers evidently came in useful as 'fillers' when other material ran short.

32 At least two-thirds of the members were field botanists first and foremost, in the opinion of Cooper ('On the advancement of local botany in the environs of London...', *Mag. Zoo. Bot.*, 2 (1838) 163–170).

33 D. Cooper, 'Details of the First Excursion made this summer by the members of the Botanical Society of London', *Mag. Nat. Hist.*, n.s. 2 (1838) 556–559; reprinted in *Proc. Bot. Soc. Lond.*, 1 (1839) 74–76.

34 G. E. Dennes to T. B. Hall, November 5th, 1838: LIV. A comparable letter of the Edinburgh society's has survived (W. Brand to N. J. Winch, May 23rd, 1836: Winch correspondence 8/51, Linnean Society of London), which shows that the function of its Local Secretaries was similar, also that it was prepared to reimburse them for any expenses incurred and that even non-members were acceptable for the office.

35 To judge from the frequent attendances of a member (Thomas Twining) who is known to have been permanently on crutches, the Society's rooms were on the ground floor. As a later General Secretary (of the BSBI) was to be an ex-singer, it is not entirely inappropriate that these should now hold the office of the Concert Artistes' Association.

36 Nevertheless 'M.B.S.L.' continued to be the recognised style officially (cf. A. Hume, *The Learned Societies and Printing Clubs of the United Kingdom* (London, 1847), p. 111).

37 Publisher's advertisement at back of M. H. Cowell's *A Floral Guide for East Kent*. The price to the public was six shillings and sixpence.

38 This arrangement, which began in November 1841, was no doubt a copy of Cooper's initiative that January in securing the permission of the Microscopical Society to publish in his *Microscopic Journal* abstracts of the papers read at its meetings – which otherwise it could not afford to put into print. Additionally he had offered "to make the journal serviceable to the views of the Society" (A. D. Michael, 'The history of the Royal Microscopical Society', *J. Roy. Micr. Soc.*, Ser. 2, 15 (1895) 1–20).

39 Watson to Hooker, January 19th, 1842: HOO. This would have been in a family tradition: Cooper's father had unsuccessfully applied for a Chair in Chemistry Applied to the Arts in the University of London in 1827.

40 Numerous tributes appeared in the scientific press, the most informative being the obituary in the *Microscopic Journal* (1842: 351–352). Though unsigned, this can have been written only by J. Forbes Young.

From Chapter 3, Watson Takes Over

1 The list of members as at March 1839 that was published in the *Proceedings* does not include Watson's name. But in the annual Report that November he is one of six people thanked for especially valuable contributions of specimens. It would have been unlike Watson to have participated to this extent without enrolling as a member. His wish to benefit from the Society's distributions of duplicates – his

reason for joining the Edinburgh society earlier – would in any case have been motive enough.

2   Watson never embraced the Irish flora in his studies because it was as yet so scantily worked. He was content to leave it to Irish botanists to produce a series of companion monographs, which opened with the *Cybele Hibernica* of David Moore and A. G. More in 1866.

3   For biographical details see especially F. N. Egerton, 'Hewett C. Watson, Great Britain's first phytogeographer', *Huntia*, 3 (1979) 87–102 and his entry on Watson in the *Dictionary of Scientific Biography* (New York, 1976), vol. 14, pp. 189–191. Professor Egerton has a full-length biography awaiting publication, which I have been privileged to read in draft.

4   Watson to J. H. Balfour, February 17th, 1871: BAL.

5   Watson to Balfour, February 25th, 1841: BAL.

6   Watson to Balfour, February 25th, 1842: BAL.

7   Babington to Balfour, May 1st, 1843: Babington, p. 292.

8   Watson to Balfour, January 13th, 1846: BAL.

9   Watson to Hooker, December 16th [1840]: HOO.

10 cf. Dennes to Hall, April 23rd, 1839, LIV: "... you must endeavour to send in as many Plants as possible at the present time. I hope to send in some thousands of Plants myself."

11 Dennes to Hall, January 2nd, 1840: LIV.

12 In his letter to Hooker (note 9, above) his estimate beforehand had been 20,000 specimens – twice as great as this.

13 Taking into account what had been distributed already, Watson reckoned the total of misnamed British specimens must have run into hundreds (Watson to Babington, January 29th, 1842: BAB).

14 See Ch. II note 20.

15 Flower, f. 21.

16 *Topographical Botany*, ed. 2 (London, 1883), p. 568. Not surprisingly, Druce (*British Plant List*, ed. 2 (Arbroath, 1928), p.v; and elsewhere) mistakenly supposed that Dennes's post was a paid one.

17 See note 5. The Edinburgh society did not switch to a regular, paid Curator till two

years later, when W. W. Evans, a young gardener, was appointed (with the title of Assistant Secretary).

18 E. Doubleday to Hooker, October 21st, 1843: HOO.

19 See Hortense S. Miller, 'The herbarium of Aylmer Bourke Lambert: notes on its acquisition, dispersal and present whereabouts', *Taxon*, 19 (1970) 489–553. The name 'Henfry' features among Mrs Miller's unidentified purchasers. Other members secured lots on their own account and presented these to the Society at the following meeting (according to the report of that in the *Annals & Magazine of Natural History*). It was presumably from this source that the Society also came by Sole's set of British Mints, its acquisition of which was reported that year (allegedly by donation from the impecunious Woodward).

20 Watson to Hooker, August 12th, 1844: HOO.

21 Watson to Hooker, September 18th [1844]: HOO. See also *The Young Shetlander ... Life and Letters of Thomas Edmondston*, ed. by his mother (Edinburgh, 1868), p. 199. Within eighteen months Edmondston met his death, off the coast of South America.

22 Watson to Babington, August 14th, 1847: BAB. Babington was able to come to a comparable arrangement with the Edinburgh society, in return for overhauling its collections in September 1844 (Babington to Duthie, August 13th, 1874: Babington, p. 372).

23 *The Geographical Distribution of British Plants* (Thames Ditton, 1843), p. 17.

24 *Phytologist*, 1 (1842) 496. The Edinburgh society included a similar idea in its original set of aims, in 1837, doubtless at Watson's inspiration too.

25 Only one copy of the circular is known to exist: among the T. B. Hall papers in Merseyside County Museums. Its date is probably 1841. A slightly modified version was appended to (at least) the 1843 Prospectus, copies of which survive in the British Museum (Natural History) and British Library.

26 The Revd Andrew Bloxam, for example, who had recently taken up the study of the Brambles under the tuition of Edwin Lees sent in 1847 "a very large supply" of specimens of the Midland species known to him, for distribution to fellow members (*Phytologist*, 3 (1848) 181–182).

27 *Phytologist*, 2 (1845) 28–29.

28 Watson to Hooker, February 24th [1843]: HOO.

29 It did once rise to the cost, though, of paying for collecting to be done for it in Britain. This was in 1840, in its reckless early years, when it followed the lead of the Edinburgh society in commissioning William Gardiner to send it material from

the Highlands (Anon., 'William Gardiner the botanist', *Chambers's Edinburgh J.*, 7 (1847) 248–251).

30 Mill to Bentham, March 14th, 1843: Bentham correspondence, Royal Botanic Gardens, Kew. Published as Letter 391 in *The Earlier Letters of John Stuart Mill 1812–1848*, ed. F. E. Mineka. Toronto & London, 1963.

31 Babington, p. 310.

32 Mary Munslow Jones, *The Lookers-Out of Worcestershire* (Worcester, 1980), pp. 97–99.

33 cf. a curt note of his to Sansom, the Society's Librarian, January 5th, 1848 (British Library Add.MSS. 45879, f.165^v): 'In any future parcel, pray attach the label by a *single* slit, at the *base*: and take care to put the *No*. . . from the London Catalogue.'

34 *Phytologist*, 3 (1848) 38.

35 Watson to Hooker, October 30th [1850]: HOO.

36 Watson to Balfour, August 18th, 1846: BAL.

37 Watson to Babington, June 16th, 1844: BAB.

38 Gray's portrait later passed to the Royal Society and now hangs just outside the library. It was reproduced in 1974 in *J. Soc. Biblphy Nat. Hist.*, 7: 38 and in 1980 in *The Royal Society Catalogue of Portraits*. Watson's is now in one of the museums at Kew. They are included in the present book as Plate 2 and the Frontispiece respectively.

39 *Phytologist* (N.S.), 6 (1863) 470. Gray, a more reliable witness, mentions portraits only of the President, Vice-President and Secretary as having been painted by subscription (Gunther, p. 220).

40 *Phytologist* (N.S.), 5 (1861) 84.

41 *Phytologist*, 3 (1849) 478–488.

42 *Phytologist*, 3 (1850) 801–811.

From Chapter 4, The Other Side

1 Commuting from beyond the suburbs, though practicable from the late 1830s, was still unusual at this period. As late as the mid-1850s only nine per cent of those journeying to work in the Capital, it was estimated, used the trains even from closer in – hardly any more than those using carriages or hansom cabs. Two out of every three still lived near enough to their place of work to travel to it on foot

(Report of Select Committee on Metropolitan Communications, *Parliamentary Papers* 1854–55, vol. 10, App. pp. 215–216). There was no Underground, nor even Victoria Station, till the 1860s.

2 Helped in particular by an influx of Irish members, following the failure to float a Botanical Society of Dublin in 1842 (on which see E. C. Nelson, *Irish Nat. J.*, 20 (1981) 282). Throughout its existence, however, the BSL succeeded in recruiting no more than thirteen Irish residents – a mere four per cent of the total.

3 G. E. Smith to J. Barton, May 9th, 1856: published in the *Naturalist*, 103 (1978) 32.

4 He had been active in the founding of the London Mechanics' Institute (now Birkbeck College) back in 1823 and was an original member of the College of Preceptors in Bloomsbury from 1846 onwards. The pure sciences were to the fore in the teaching at such institutions.

5 R. D. Radcliffe, *A Memoir of Thomas Glazebrook Rylands* . . . (privately printed, 1901), p. 36.

6 Morris Berman, *Social Change and Scientific Organization: the Royal Institution, 1799–1844* (London, 1978), pp. 116, 123.

7 A. H. Hassall, *The Narrative of a Busy Life* (London, 1893), p. 43.

8 See Gray's autobiographical fragments, reproduced in *J. Soc. Biblphy Nat. Hist.*, 7 (1974) 69–71.

9 The six were Isaac Brown, Corder, Greenwood, G. S. Gibson, J. M. Gibson and Wallis.

10 According to Edward Newman (*A History of British Ferns*, ed. 3 (London, 1854), p. 126), who evidently knew him personally.

11 A. Thackray, 'Natural knowledge in cultural context: the Manchester model', *Amer. Hist. Rev.*, 791 (1974) 672–709.

12 Obituary in *Nottingham Daily Express*, July 20th, 1899.

13 On Druce's technique see R. W. Butcher, *Proceedings*, 1 (1955) 396.

14 Gage, pp. 146–7. One member's subscriptions were overdue for the astonishing total of 23 years.

15 Though the figures are shaky from 1849 onwards, for deaths (at least) were apparently under-reported.

16 I have set out the evidence for this identification in Newsletter no. 12 (1981, pp. 8–9) of the Society for the Bibliography of Natural History.

From Chapter 5, Collapse

1 Syme, *Entomologists' Weekly Intelligencer*, February 26th, 1859, pp. 172; Irvine, *Phytologist* (N.S.), 6 (1863) 470. Conceivably Irvine's use of "split" is only metaphorical.

2 *Phytologist*, 3 (1850) 801.

3 This was also a much simpler task in itself; for Watson had taken the view that the Foreign Collection was too inadequately named and arranged for material of particular species to be selected from it, so instead distribution was done 'in the lump', in packets according to country or collector (Watson to Hooker, October 9th, 1847: HOO). Watson had tried to simplify matters further by urging the Society to confine itself to the European flora only, but this had not proved acceptable.

4 Watson to Hooker, October 30th [1850]: HOO.

5 *Phytologist*, 5 (1854) 95–96.

6 *Phytologist* (N.S.), 1 (1855) 117.

7 Syme to Hooker, February 28th, 1855: HOO.

8 Syme to Wilson, March 18th, 1856: WIL.

9 *Phytologist* (N.S.), 5 (1851) 84.

10 For example, Syme to Hooker, November 20th, 1856: HOO.

11 October 2nd, 1856: BAB.

12 February 17th, 1871: BAL.

13 The paper was used for copying out Miss E. A. Warren's MS Flora of Cornwall, which had been sent to him on loan. The transcript is now among the (unnumbered) Watson MSS at Kew. The text of the circular was reproduced by its discoverer, J. E. Lousley, in *Proceedings*, 5 (1963) 119–120.

14 January 19th, 1857: HOO.

15 BM-B Library, shelfmark 58.00p5, no. 22. Several of the collections donated to the Society can be identified from their descriptions despite the absence of collectors' names.

16 *Topographical Botany*, ed. 2, p. 557.

17 Druce (quoting Baker), *History of the Club* (Oxford, 1911), p. [2]; offprinted from

preface to Report Vol. 2. Baker soon after borrowed from Hardy the original set of British Mints formed by William Sole, which had been presented to the Society in 1842. Probably it was just this one collection that Baker had in mind when recalling Hardy as a purchaser at the auction.

18 *Phytologist* (N.S.), 2 (1857) 143; *Gardeners' Chronicle* (1857) 648. Brocas subsequently presented himself as the natural successor to Pamplin, when the latter retired from dealing in 1862 (J. Sim to Wilson, November/December 1864: WIL).

19 Watson to Baker, January 19th, 1866: published in *Proceedings*, 4 (1962) 410–412. The original of this key letter was found by the late J. E. Lousley tipped into a book he had acquired. It remains in the possession of his widow.

20 *J. Bot.*, 60 (1922) 364. Early this century James Britten seems to have tried hard to establish the facts, but his investigations proved inconclusive.

21 Note 17; and reported in 1917 Report, pp. 12–13.

22 BM-G, Miscellaneous Papers, f. 9: published in Gunther, p. 220. Gray was further in error in adding that the Society presented his portrait to the Royal Society. In fact the Royal Botanic Society had it first.

23 J. de C. Sowerby to Mons. Kralik, August 2nd, 1859: British Library, Egerton MSS. 2851, f. 180.

24 J. R. Norman, *Squire: Memories of Charles Davies Sherborn* (London 1944), p. 121.

25 A few years before the Society's founding the infant British Association gave serious thought to launching botanical enquiries on topics which might best be studied "by being pursued by persons in various parts of the Kingdom". They included determining the distribution of certain plants according to the soils and strata on which they grow and compiling a list of species arguably introduced by man. However, when Henslow, the Professor of Botany at Cambridge, proved unwilling to act as organiser, the idea was quickly dropped (W. V. Vernon Harcourt to W. Whewell, October 17th, 1831; Whewell to Harcourt, December 19th, 1831: published in *Gentlemen of Science: Early Correspondence of the British Association for the Advancement of Science*, ed. Jack Morrell & Arnold Thackray (London, 1984), pp. 88, 116).

26 Strictly speaking there were only 112 Watsonian vice-counties, for he carved up only Great Britain (in which he included the Isle of Man), leaving Ireland on one side for the time being and disregarding the Channel Isles as too alien geographically. Forty Irish vice-counties were devised by other hands later, and it also became conventional to treat the Channel Isles ('Sarnia') as one further.

27 *Phytologist*, 3 (1850) 949.

28 Though Moore's experiment was not reported in print till 1866 (in *Cybele Hibernica*,

ed. 1, p. 240), there is a description and watercolour of the plant in question dated "24 Aug. 1853" in Moore's hand (det. E. C. Nelson) in a collection of drawings of critical British plants formed by the Revd G. E. Smith now in BM-B. H. G. Baker (*Proceedings*, 1 (1954) 135) has described the experiment as an exercise in genecology, but D. A. Webb (in litt. to the author, 1984) considers that is going too far.

From Chapter 6, Interlude at Thirsk

1 'H.B.' [Harriet Beisly?], *Phytologist* (N.S.), 2 (1857) 246.

2 J. H. Davies to Wilson, November 10th, 1857: WIL.

3 Syme, for his part, had chosen this juncture to take up entomology more seriously. In 1858 he was elected to the Council of the Entomological Society of London – only two years after joining.

4 Baker to Druce, November 14th, 1917: published in 1917 Report, pp. 12–13.

5 Baker to Wilson, August 23rd, 1854: WIL. Newman had already tried Watson, if not others.

6 *The Flowering Plants and Ferns of Great Britain; an Attempt to Classify them according to their Geognostic Relations* (London, 1855). This was an extended version of a paper he had read before the British Association at Glasgow.

7 Baker to Wilson, October 11th, 1857: WIL.

8 *Phytologist* (N.S.), 4 (1860) 55.

9 This was certainly the view taken by Druce and by the Botanical Exchange Club subsequently – which celebrated 1936 as its centenary, therefore.

10 *Topographical Botany*, ed. 2, p. 605.

11 Watson to Hooker, February 13th, 1861: HOO; Watson to G. A. Walker Arnott, February 18th, 1861: Walker Arnott correspondence, BM-B.

12 Watson to Babington, February 16th, 1861: BAB.

13 ibid.

14 Watson to Baker, November 12th [1873]: BAK.

15 W. Carruthers to Babington, November 27th, 1862: pasted into back of Cambridge Botany School Herbarium copy of *Journal of Botany*, vol. 1.

16 Watson to Baker, January 14th [1871]: BAK.

17 The fullest of several accounts of this society is that in *J. Bot.*, 2 (1864) 287–288.

18 See J. Ramsbottom, 'The Society of Amateur Botanists and the Quekett Microscopical Club', *J. Quekett Micr. Club*, Ser. 2, 16 (1931) 215–230.

19 T. J. Foggitt, 'Annals of the B.E.C. II Recollections of the Thirsk Botanical Society', 1932 Report, pp. 289–297. It is suggestive that Foggitt, in misremembering the name of the Club, should have given it one so reminiscent of its London ancestor.

20 *Naturalist*, 1 (1865) 42.

21 J. D. Hooker to Baker, November 20th, 1865: J. G. Baker correspondence, Kew.

22 Watson to J. D. Hooker, January 13th, 1866: HOO.

23 Watson to [Baker], December 16th [1865]: BM-B autograph collection.

24 Watson to Baker, January 19th, 1866: published in *Proceedings*, 4 (1962) 410–412.

From Chapter 7, The Years of Obscurity

1 The terms of the inheritance required him to take the maternal surname of Boswell. Mindful of the confusion this meant for taxonomy, he compromised at first with Boswell-Syme; but the lawyers eventually objected and in 1875 this had to be abandoned. For the sake of clarity Syme is employed through this book.

2 Watson to Baker, October 28th, 1875: BAK.

3 *Supplement to the Compendium of Cybele Britannica* (Thames Ditton, 1872), pp. 211–213: 'Letter to the Members of the Botanical Exchange Club'.

4 Trimen to Duthie, April 28th, 1874: Duthie letters 2/221, Kew.

5 Flower, ff. 27–28.

6 Flower, f. 34. Flower asserted Hardwicke was bankrupted by it, but that cannot be so, for the latter's business was sold on his death as a going concern (Mary P. English in litt. to the author, 1984).

7 Druce, 1924 Report, p. 530.

8 *J. Bot.*, 17 (1879) 160.

9 In the two accounts Druce much later gave of this affair (1924 Report, p. 530 and

1928 Report, p. 709), as related to him by Bailey, the impression is conveyed that it took place around 1897. All the other evidence, however (including Bailey to Druce, September 4th, 1895 and September 1st, 1898: DR 29), points to 1879–80.

10 No surviving copy of either this or the (putative) second circular has been traced. All we have is Babington's reply to the letter (Babington, p. 393).

11 Though he seems to have had potential as a taxonomist. He became widely known, it is said, as an expert on trademarks, which he studied with scientific minuteness and precision (F. E. Weiss, *North West Nat.*, 5 (1930) 81–86).

12 But Druce (1915 circular, p. 2; 1928 Report, p. 709) surely exaggerates in depicting the Watson Club as substantially the creation of a breakaway BEC faction led by Linton.

13 Druce, 1903 Report, p. 4.

14 W. D. Foster, 'The history of the Moss Exchange Club', *Bull. Brit. Bryol. Soc.*, no. 33 (1979) 19–26.

15 Bailey to Druce, December 29th, 1892: DR 29.

16 Linton to Druce, January 27th, 1887: DR 4.

17 White to Druce, April 13th, 1906: DR 11.

18 Bailey to Druce, May 14th, 1888: DR 29.

19 E. S. Marshall to Druce, February 15th, 1902: DR 17.

20 A. Somerville to Druce, October 28th, 1903: DR 17.

21 J. W. White complained (September 3rd, 1893: THO) that most of the bramble specimens he received through the BEC were in such very bad condition that nine out of ten had to be burnt.

22 Druce, 1916 circular, p. 2. Bailey circulated a set of accounts, at last bringing the situation to members' notice, as a hectographed appendage to the 1895 Report.

23 In just the same way, though on a more modest scale, the affluent Mrs T. A. Cotton rescued the Watson Club from occasional deficits over the years.

24 Bailey to Druce, January 27th, 1898: DR 29.

25 cf. Miss C. E. Palmer to Druce, December 27th, 1899: DR 19.

26 Renamed in 1870 the Botanical Record Club: Phanerogamic and Cryptogamic.

27 When Watson's '*Topographical Botany*' herbarium passed to Kew on his death, the Record Club was anxious that its specimens should be stored alongside that, in order to emphasise that they represented a continuation of his work. But the Kew authorities found this impracticable and in 1884 suggested that the voucher collection would be better transferred to the British Museum (Natural History). To this the Club, rather reluctantly, consented (Bailey to D. Oliver, October 25th, 1884: Botany Dept. correspondence, BM-G). The sheets, 1340 in number, are consequently now incorporated in the Museum's British herbarium.

28 Bailey to Druce, February 12th, 1891 and June 30th, 1892: DR 29.

29 Ward to Linton, October 14th, 1890: Linton autographs, BM-B.

30 Anon., 'A new British Flora', *J. Bot.*, 36 (1898) 94–96.

31 W. West to A. Bennett, June 7th, 1898, quoted in Bennett to Linton, February 5th, 1912: Linton autographs, BM-B. The rival was Druce.

From Chapter 8, Civil War

1 The analysis of the replies is preserved among Druce's papers (DR red box). Bailey sent out a hectographed letter, dated December 8th, reporting the results.

2 Marshall to Britten, April 21st, 1907: British Herbarium correspondence, BM-B.

3 Marshall to W. H. Beeby, February 13th, 1909: Beeby correspondence (in possession of Mrs D. W. Lousley).

4 In the year after Druce's election from 37 per cent to 29 per cent. In two more years the figure was down to 18 per cent.

5 H. J. Riddelsdell to Druce, February 27th, 1902: DR 41.

6 A. B. Rendle, *Obituary Notices of Fellows of the Royal Society* (London, 1932), pp. 12–14.

7 Druce to Linton, January (or June) 28th, 1907: Linton autographs, BM-B. Linton was not won over.

8 Druce to Marshall, June 29th [1915]: DR red box.

9 Draft letter to W. C. Barton, June 26th [1915]: ibid.

10 Bailey to Druce, February 20th, 1911: DR 29.

11 Druce to P. M. Hall, February 27th, 1928: Hall papers, BM-B. The feud started as a result of Britten altering a manuscript of Druce's without reference to him.

12 Riddelsdell to Druce, February 27th, 1902: DR 41.

13 Marshall to Druce, February 29th [1915]: DR red box.

14 The more unappealing for being so largely concerned with non-British species quite outside the concern of the Club – for example, lists of 136 new combinations in the 1913 Report and of over 500 in that for 1917. Druce (1915 circular, p. 4) justified the publication of these on the ground that the Report was thereby "brought before botanical authorities" and might gain the Club fresh members.

15 Draft letter to J. E. Little, July 18th, 1915: DR red box. The blocks for the 1912 Report were paid for by Miss Ida Hayward.

16 cf. Marshall to Druce, August 9th, 1911: DR 19.

17 Marshall to Druce, March 5th, 1910 and February 22nd, 1911: DR 19.

18 Druce, 1930 Report, p. 326.

19 Moss to Rendle, May 25th, 1915: Botany Dept. correspondence, BM-G.

20 Druce, 1915 circular, p. 6.

21 He was presumably weekending at Longleat.

22 Druce to Rendle, June 5th [1915]: Botany Dept. correspondence (under 'Bot. Exch. Club'), BM-G.

23 Marshall to Rendle, June 11 [1915]: ibid.

24 A copy of this circular, all the replies received to it, and the subsequent correspondence generated are to be found among the Druce papers crammed into a small red card-index box labelled, in Druce's hand, 'The Row'. This makes it unnecessary to cite the sources individually.

25 Copies of both Moss's circulars exist in BM-B and DR. The former were donated by the Cambridge Botany School in 1931.

26 Marshall to H. Stuart Thompson, November 13th, 1915: THO.

27 *Manchester Guardian*, March 2nd, 1932.

28 *Pharmaceutical J.*, March 4th, 1932.

29 A copy of this second circular has survived in the Society's archives. It bears no date, but a letter from R. H. Corstorphine to Druce shows that it was in draft by October 12th. Like all the other circulars, it is headed 'private and confidential'

(and, miraculously, not so much as a hint of their existence did ever leak into the public literature.)

30 Moss to Thompson, February 5th, 1917: THO.

From Chapter 9, King Druce

1 cf. C. E. Salmon to Rendle, June 18th, 1915 (Botany Dept. correspondence, BM-G): "It is very difficult to find out how many disapprove of the Secretary's actions or approve! If only a meeting could be called we could find out – but then the Secretary would call the meeting and be present!"

2 No such body is known to Dr J. Sheail, the leading authority on the history of the British nature conservation movement (in litt. to the author, 1983). The name, however, is repeated by Druce in his entries in *Who's Who*. Presumably it was an early variant in use for the Society for the Promotion of Nature Reserves, which was then only three years old and perhaps still lacked a settled title. Druce was certainly a member of the Executive of the SPNR.

3 1915 circular, p. 5.

4 Bailey to Druce, March 16th, 1914: DR 29.

5 Draft circular, [June 1915]: DR red box. But in the printed version of this (p. 5) he wrote that the project had to be *given up* – a telltale alteration.

6 Hiern to Druce, March 6th, 1915: DR red box.

7 Marshall to Druce, June 17th, 1914: DR 19.

8 1923 Report, pp. 245–249. This gives the lie to R. W. Butcher's allegation (*Proceedings*, 1 (1955) 396) that there were no field meetings during the Druce period.

9 P. D. Sell, oral information.

10 to Druce, May 16th, 1888: DR 19.

11 E. C. Nelson, *Western Nat.*, 6 (1977) 63.

12 W. Whitwell to Linton, August 9th, 1905: Linton autographs, BM-B.

13 H. W. Daltry to Druce, July 17th, 1915: DR red box.

14 Lt-Col. J. Codrington, oral information.

15 For this same reason he seems to have made a special play for the plutocracy too:

Sir Jeremiah Colman (of Colman's Mustard), Reginald Cory (of Cory Brothers, the coal distributors) and Lord Melchett (of ICI) must have been obvious 'easy touches'. Also in the membership list of 1929 are a Watney, a Whitbread and three Rothschilds.

16 Druce, *J. Northants. Nat. Hist. Soc. & F.C.*, 19 (1918) 131–142; J. Wake, *Northants. Past & Present*, 1²(1952) 39–52.

17 *Nature*, March 19th, 1932.

18 Actually only one in ten were titled in the strict sense. But Druce is unlikely to have made any mental distinction between these and those styled 'Hon.' (or 'Rt. Hon.').

19 Pearsall to P. M. Hall, March 7th, 1932: PE.

20 Corstorphine to Druce, November 16th, 1920: DR 14.

21 Riddelsdell to Druce, August 30th, 1929: Riddelsdell *Rubus* MSS, BM-B.

22 e. g. Pearsall to Druce, April 6th, 1917: DR 8.

23 The *Comital Flora* may have had its origin as the 'Ecological Flora of the British Isles' which Druce announced in the *Pharmaceutical Journal* in 1900 that he had in preparation. In this he aimed to show "more particulars as to the exact place of growth, altitude and distribution than is given in the usual text-books". It was thus Watsonian in conception rather than ecological in the sense in which Moss and Tansley understood that term.

24 Andrew Young, *A Prospect of Flowers* (London, 1945), p. 62.

25 This was as early as 1919, suggesting that the pursuit of 'royal' status had been among his foremost priorities.

From Chapter 10, Return to Democracy

1 His old Oxford friend, F. A. Bellamy, the BEC's Honorary Auditor for many years, was suggested for this task or, failing him, Pearsall or Horwood. In the event, though, nothing ever appeared in print and presumably the sum went to the residuary legatees.

2 Pearsall to Hall, March 23rd, 1933: PE.

3 The negotiations were not made easier by the fact that Druce and the new Professor of Botany, Tansley, detested one another.

4 Wilmott, General Committee minutes, October 22nd, 1941.

5 Pearsall to Hall, June 27th, 1933: PE.

6 The SPNR invested its share, which promptly trebled its annual income (John Sheail, *Nature in Trust* (London, 1976), p. 66).

7 Pearsall to Hall, March 3rd, 1932: PE.

8 Nigel West, *MI5* (London, 1983), p. 47.

9 1932 Report, p. 17.

10 Pearsall to Chapple, June 20th, 1932: DR 39.

11 Pearsall to Druce, April 7th, 1931: DR 19.

12 Mrs Foggitt to Druce, April 7th, 1931: DR 5.

13 Pearsall to Hall, March 6th, 1932: PE.

14 In the event only seventeen places were filled, for James Groves was found to have sent in his resignation shortly before the meeting and so (to general relief) was ineligible.

15 Pearsall would have preferred Ramsbottom as Chairman, on the grounds that he "knows everybody and *hears everything*". Baker he thought much too quiet and over-dependent on advice (Pearsall to Hall, November 8th, 1932: PE).

16 Lousley, *Proceedings*, 4 (1961) 352.

17 Pearsall to Chapple, June 20th, 1932: DR 39.

18 Mrs Foggitt to Chapple, August 1st, 1935: DR 27.

19 Lousley to Thompson, October 9th, 1935: THO.

20 The membership of the Horticultural Society similarly fell in 1931 and 1932, only to recover in 1933 and steadily increase thereafter (Harold R. Fletcher, *The Story of the Royal Horticultural Society 1804–1968* (London, 1969), p. 336).

21 With a nice historical touch, the first field meeting in that inaugural 1934 programme was at Woking, "thereby repeating the Society's first excursion of ninety-seven years ago" (*Gardeners' Chronicle*, May 26th, 1934, p. 339).

22 Pearsall to Chapple, December 20th, 1932: DR 12.

23 Pearsall to Thompson, January 10th, 1934: THO. Even so there were complaints that at 3/6d the tickets were too expensive (Chapple to Hall, September 9th, 1937: BSBI (under 'BEC Meetings').

24 Mrs Foggitt to Chapple, November 26th, 1935: DR 27.

25 Unfilled at any rate by the Society. A. R. Clapham's checklist of 1946, published in the *Journal of Ecology* in connection with the *Biological Flora of the British Isles*, served as a very useful holding measure till J. E. Dandy's definitive volume eventually appeared in 1958 (see p. 152).

26 She is said to have been the first woman to fly in an airship or aeroplane. Of these and other exploits, related in her *Memories of Land and Sky* (London, 1928), one of five books written under her maiden name of Gertrude Bacon, her botanical friends remained largely unaware. In earlier days she lectured to schools on her experiences. Sir Rupert Hart-Davis, in his autobiography, *The Arms of Time* (London, 1979; p. 101), mentions hearing her in this capacity when a schoolboy around 1920.

27 Wilmott to Pearsall, August 2nd, 1935 and January 31st, 1936: copies in Wilmott correspondence, BM-B.

28 Wilmott to Hall, January 20th, 1939: BSBI (under 'BEC Meetings').

29 A very shy retired solicitor, Francis Druce was another Bailey: a born treasurer, happy to serve in that capacity more than the one society at once. The Linnean Society shared him with the BEC.

30 Pearsall to Hall, April 24th, 1932: BSBI. Pugsley's antipathy to Druce went back to early in the century, when he along with Moss and Britten sought to compel him to resign from the Linnean Society by having him censured for using his Fellowship for trade purposes (Druce to Ramsbottom, April 14th, 1931: British Herbarium correspondence, BM-B; Bennett to Linton, April 12th, 1915: Linton autographs, BM-B).

31 Watson Club circular, May 7th, 1934 (copies in THO and BSBI).

32 Thompson to Pearsall, April 24th, 1934 – quoted in Pearsall to Thompson, April 28th: THO. The leading Watson members also hated the idea of being swallowed up in a body in which "nine-tenths of the members are not botanists and have never contributed to the exchange" (Pugsley to Thompson, January 19th, 1934: in Thompson's copy of Watson Club Report, Vol. 4, Botany/Zoology Library, Bristol University).

33 Watson Club circular, as above. This name had in fact been suggested many years earlier by James Groves (to Druce, January 19th, 1914: DR 21). It has also been generally overlooked that Druce himself at least once employed it in print, as a variant on 'BEC', in his *Flora of Buckinghamshire* in 1926 (p. xcvii). He may have been in the habit of using it in conversation, for certainly it sounded more impressive.

From Chapter 11, The Great Efflorescence

1 Known to everyone as 'Ted'. The unusual first name was inherited from a botanist ancestor, on whose Berkshire work he was to publish an interesting paper (*Proceedings*, 5 (1964) 203–209).

2 Indeed he did talk seriously at one time of following botany for a living (cf. Lousley to H. Stuart Thompson, March 1st, 1928: THO).

3 Lousley, in litt. to the author, April 1972. It remains difficult to understand how this could have happened had Rule 2 (iv), introduced in 1935, been operating efficiently.

4 Treasurer's report to General Committee, March 25th, 1942.

5 The hand was that of D. P. Young, the final person to serve as Distributor. Strictly speaking the Exchange section was merely placed in suspension; to date, however, that suspension has been permanent.

6 The first conference, in 1948, received the impressive accolade of no less than a fourth leader in *The Times*. For this the Society had to thank the fact that its anonymous author, I. A. Williams, combined writing for that paper with long-standing prominence as a field botanist (being particularly renowned for his work on *Bromus*). Hardly less impressively, *The Times* of the scientific world, *Nature*, proved willing to publish the detailed reports which Lousley conscientiously sent in on each year's Exhibition Meeting. These twin publicity breakthroughs seemed to signal the arrival of the Society on to an altogether higher plane of reputability.

7 This had been proposed (by A. E. Wade, a near-teenage recruit himself) some three or four years earlier, but initially rejected because of fears that schoolchildren could not be relied on to observe the Rules and otherwise behave responsibly.

8 Miss Campbell to H. T. Baker, January 21st, 1939: BSBI (under 'BEC Meetings'). She rejoined in 1943.

9 Other than in the capacity of annual Distributor, that is.

10 Warburg to Miss Campbell, August 30th, 1946: Campbell papers, BM-B.

11 One of the two dissenters was R. A. Graham, who later carried his objection to the point of staging an exhibit which showed that the name was also that of a brand of tinned meat.

12 Lousley to Swann, July 1st, 1955: BSBI.

13 In 1953, for example, 15 per cent of the 92 new members were identifiably secured by the activities of this committee.

14 It was customary by this time to use an initial capital when referring expressly to those officially so designated in the Rules. Not all office-holders are or ever have been Officers in this strict sense.

15 Report of the Honoraria Sub-committee, October 1949. The honorarium paid to Chapple in the 1930s appears to have been discontinued after the War.

16 Treasurer's memorandum, August 10th, 1947: BSBI.

From Chapter 12, Full Steam Ahead

1 For full details see my paper, 'The Botanical Society's regional structure', *Proceedings*, 4 (1961) 155–159.

2 The question of reimbursing representatives for the expenses incurred in attending these became a matter of contention. Lousley, who was hotly opposed to doing so in full, pointed out that just a single enquiry could give rise to a claim equivalent in cost to an extra eight pages for *Watsonia*. In the end it was agreed that a maximum of £10 could be paid just in cases where a member was asked to attend an enquiry at a particularly great distance from home.

3 A detailed account of the numerous and complex developments which took place on the national conservation front in these and subsequent years is to be found in Chapter Eight of John Sheail's *Nature in Trust*. For fuller details of the BSBI's activities specifically E. Milne-Redhead's 1970 Presidential Address (*Watsonia*, 8 (1971) 195–203) should be consulted.

4 This step cannot have been reassuring to certain influential members who distrusted Lousley as an incipient Druce. With a view to containing his power, they took the lead in seeking as the new President a particularly authoritative figure, to serve as a counterweight (N. Douglas Simpson, oral information). Unfortunately for this plan, Canon C. E. Raven proved able to devote little time to the office.

5 J. G. Dony, in litt. to the author, 1984.

6 London, 1953.

7 T. G. Tutin, in litt. to the author, 1980.

8 As a consequence of the needs of the Allied Air Forces during the war a system of navigational coordinates employing metric system units with an arbitrary origin just to the south-west of the Scilly Isles had been established. This network, entitled the Ordnance Survey National Grid, was now printed on the one inch to the mile (and larger-scale) maps published by the Survey and made it possible to locate records within areas 100 km square, 10 km square or even 1 km square.

9 One difficulty the committee early had to confront was that the National Grid covered the whole of England, Wales and Scotland but only the eastern half of Ireland. D. A. Webb elegantly solved the mathematical problem of extending the Grid to the rest of Ireland, and transferred it to copies of the ½-inch to the mile maps that were the largest scale then available. At a much later date the Irish Ordnance Survey adopted a grid system of its own with a different origin.

10 Some further small grants were received in addition: £50 from the British Ecological Society, an annual £25 from the Royal Irish Academy (in aid of field work in Ireland), and a token £50 yearly from the BSBI itself throughout the entire ten years of the Scheme.

11 Powers-Samas were subsequently able to interest a then leading firm of women's hosiery manufacturers, Kayser-Bondor, in utilising the system for mapping the sales of nylon stockings. Ironically, the latter firm was later to be bankrupted by its computer.

12 A fuller account of the genesis of the Scheme, including the methods employed, is to be found in a paper contributed by Walters to *Proceedings* (1 (1954) 121–130). The annual reports on the Scheme, also published in *Proceedings*, and the introductory chapters to the *Atlas of the British Flora* amply document the rest of the story.

13 Though the equipment soon became outdated. The punched-card sorter and tabulator used in the Scheme now grace the world's first computer museum in Boston, Massachusetts (S. M. Walters, *New Phytol.*, 98 (1984) 9).

From Chapter 13, Into the Future

1 An amateur/professional alternation had actually been intended earlier, starting with the election of an amateur in 1951 in the person of Canon Raven; but that first attempt had proved abortive.

2 The previous doubling of the membership was accomplished, by contrast, in only five and a half years. Admittedly, though, that was an easier task from so much smaller a base.

3 A detailed account of the Teesdale battle forms Chapter 4 of Roy Gregory's volume of case-studies, *The Price of Amenity: Five Studies in Conservation and Goverrtment* (London, 1971).

4 Report to Council by the Development and Rules Committee on the Proposals for Extending the Membership and Activities of the Society, September 1965.

5 The British Rose Survey of 1953-4, organised by P. C. Sylvester-Bradley, had been an isolated (and by then largely forgotten) pioneer attempt at one of these.

6 D. H. Kent, oral information.

7 All through the years Praeger, the foremost embodiment of Irish field botany, conspicuously stood aloof from the BEC. As he was an Ulster Protestant by background, this is hardly likely to have been because of any animosity politically. A more probable reason was his distaste for Druce. "Good botanist, but puffed up with conceit" was the terse verdict he was once heard to pass on him (Mrs N. F. McMillan, oral information).

APPENDIX ONE

Membership Figures of the Botanical Society of London

AS AT NOVEMBER 29TH EACH YEAR

	additions (incl. re-elections)	losses	net gain	total	total subscribing[1]
1836				38	38
1837	37	10	27	65	?
1838	47[2]	12	35[2]	100	68
1839	26	8	18	118	88
1840	23	13	10	128	98[E]
1841	20	6	14	142	112[E]
1842	13	3	10	152	125[E]
1843	13	6	7	159	133[E]
1844	17	3	14	173	148
1845	16	7	9	182	158
1846	27	8	19	201	177
1847	32	11	21	222	200
1848	23	8	15	237	215
1849	15	(3)	(12)	(249)	227
1850	13	(7)	(6)	(255)	233[E]
1851	17	(3)	(14)	(269)	247[E]

1852 onwards: figures were never disclosed

(Suspect figures in brackets   E = estimate)

[1] Resident and Corresponding Members only: Foreign as well as Honorary Members did not pay.

[2] Including 30 Honorary Members (category instituted April 1838).

APPENDIX TWO

The Members of the Botanical Society of London

The BSL only ever published one list of members, and that just two and a half years after its founding; moreover, it is of very limited value, for no addresses are given for the London members, making some of these unidentifiable. Fortunately it has proved possible to supplement this very extensively from other sources: the names (and, in one journal, the towns of residence) of those elected meeting by meeting throughout the middle period; the minuted lists of subscribers to the two portrait funds; lecturers, discussants and donors receiving mention in meeting reports; the surviving annual reports and prospectuses for certain of the early years; and labels in various herbaria, most notably Watson's, of specimens acquired demonstrably or putatively via the BSL Distributions.

The meeting reports, however, are not without traps, for by no means all speakers or donors would seem to have been members. Similarly non-members occasionally enriched the exchanges (especially that of 1848) but without being distinguished as such in the annual lists of acknowledgments. Herbarium labels, too, do not always make clear that the named collector was not necessarily the distributor of the specimens — which may also have come, without reference to the individuals concerned, through donation by a third party, most commonly the Botanical Society of Edinburgh. There is the yet further problem that some whose names recur repeatedly in accounts of the BSL's activities — the bryologist Dr Thomas Taylor, for example — are known to have refrained from joining.

For all these reasons an element of uncertainty attaches to many of the names in the list that follows. These are accordingly distinguished by being *placed in square brackets*. The survivors of a rigorous sifting, they have been included because in their cases the balance of probability seems to lie in favour of membership rather than the reverse. The evidence for this, and the grounds for the identifications arrived at in debatable cases, is set out on the individual cards to be deposited in due course in the Botany Library of the British Museum (Natural History).

In one or two instances initials have been found to be wrong or names misspelt in the published or even minuted record. For this, Dennes' handwriting is almost certainly to blame in part. Some of those unidentified, therefore, may have defeated all enquiry for the very simple reason that a different person was intended.

Only about half of those in the list feature in standard works of reference, despite the exceptionally wide coverage of the biographical dictionaries that exist for botany. For identifying the more obscure, recourse has mainly been had to the 1841 and 1851 Censuses and the Somerset House calendar of wills and administrations; in some cases local directories have also proved useful.

Where the year of birth has had to be calculated from the known age at death, the example of Britten and Boulger's *Biographical Index of British and Irish Botanists* has been followed by according it a '?' instead of the inherently vaguer 'c'. Dates thus queried are to be read as being two years out at the most.

Qualifications, towns of residence and occupations are those during the known or presumed period of membership only, in so far as it has been possible to establish this.

The non-paying categories of Honorary and Foreign Members have been excluded.

ABBREVIATIONS USED

(C)	Council member.
(LS)	Local Secretary.
*	Participant in the Distribution.
'onwards'	Indicates membership which continued at least until 1853 and so, presumptively, up to the Society's demise.
fl	Indicates precise year(s) of membership not known.
?	Queried membership dates are those that are probable but uncertain. A query is also employed where the year of election is known but not the subsequent period of membership.

Allman, William, M.D. (1776–1846), Dublin; professor of botany	fl. 1846
*Anderson, John (1812–post 1880), Richmond, Surrey; general practitioner	Nov. 1837–?1841 (C)
*Andrews, William (1802–1880), Dublin; army agent	fl. 1843–8 (LS)
[Ansell, John (d. 1847?), Hertford; nurseryman]	fl. 1846
Arnold, —, London (?John Roger Arnold, chronometer maker)	Dec. 1836–?
*Atkins, Mrs Anna (1797–1871), Halstead, Kent; novelist, wife of landowner	1839/40–53 or later
Attersoll, Miss Maria (1793–1877), Weymouth	1839–?
*Atwood, Miss Martha Maria (1810?–1880), Bristol	1852/3 onwards
Ayres, Philip Burnard, M.D. (1814–1863), Thame, then London; physician	Dec. 1844–9 or later (C)

*Baber, Revd Harry (1817–1892),   Nov. 1838–?43
  Cambridge University and Ely; student, then curate
*Baker, John Gilbert (1834–1920),   Sept. 1851 onwards
  Thirsk; wholesale draper and grocer
*Ball, John (1818–1889),   May 1846 onwards (C)
  Dublin and London; politician
*Banker, John (1815?–1854/66),   fl. 1852 onwards
  Devonport; messenger, H.M. dockyard
*Barham, Frederic (1809–1878),   fl. 1844–51 (C)
  London; general practitioner
Barlow, Thomas Worthington (1823–1856),   Mar. 1845–?48
  London; law student
*Barnard, Miss Alicia Mildred (1825–1911),   Sept. 1848 onwards
  Norwich and Royston; governess?
Barnard, Miss Frances Kinderley (1818?–1893),   fl. 1842
  Norwich
*Barnard, Francis (1823–1912),   fl. 1851–3
  Great Yarmouth; chemist and druggist
Barnes, James (1806–1877),   Mar. 1847–?
  Bicton, Devon; gardener
*Baxter, William (1787–1871),   Jan. 1837–?43 (LS)
  Oxford; botanic garden curator
*Bean, William (1817–1864),   Nov. 1849 onwards
  Liverpool; clerk, H.M. Customs
Beardsley, Amos (1822–1900),   Nov. 1847–?
  Heanor, Derbyshire; general practitioner
*Beesley, Thomas (1818–1896),   fl. 1841–6
  Banbury, then Chipping Norton; chemist and druggist
*Beever, Miss Mary (1802–1883),   fl. 1841–9
  Coniston; independent
Bell, William   fl. 1837
*Bentall, Thomas (1820?–1875),   fl. 1844–9
  Halstead, Essex; paper manufacturer, then farmer
Berry, Elihu (1812–1869),   Mar. 1849–?
  Ardsley, Yorks.; gardener
Betts, George Harvey (1809–1875),   Sept. 1838–? (LS)
  Coventry, then Watford; general practitioner
Biden, William Downing (1820?–post 1862),   fl. 1845–7
  Kingston upon Thames; actuary and estate agent
*Bidwell, Henry, M.D. (1816–1868),   1840/1–?50 (LS)
  Albrighton, Salop; physician

*Bigge, Revd John Frederic (1814–1885),   July 1846–?
   Ovingham, then Stamfordham, Northumb.;
   parish priest
*Bladon, James (1794–1874),   Mar. 1846–9 or later
   Pontypool; cabinet-maker
Blount, John Hillier (1822–1866),   June 1848–?
   Birmingham; general practitioner
*Bloxam, Revd Andrew (1801–1878),   1839–48 or later
   Twycross, Leics.; parish priest
Blyth, James Davis   Feb. 1847–?
*Bodenham, Thomas (1804–1873),   Dec. 1837–?45; Feb. 1847–? (LS)
   Shrewsbury; independent
[Bond, Frederick (1811–1889),   fl. 1843
   London; independent]
Booth, William Beattie (1804?–1874),   fl. 1846; Feb. 1849–?
   Carclew, Cornwall; gardener
*Bossey, Francis, M.D. (1809–1904),   Nov. 1836–48 or later
   Woolwich; physician   (C; *Vice-Pres.*)
[Brady, George Stewardson (1832–1921),   fl. 1853
   Gateshead; general practitioner]
Branfill, Miss Eliza Jemima Mary (1822–1907),   Nov. 1836–40?
   Upminster, Essex; schoolgirl
Bransby, Revd John (1783?–1857),   Aug. 1845–?
   King's Lynn; schoolmaster
*Bree, Revd William Thomas (1786–1863),   1839/40–42?
   Allesley, Warwicks.; parish priest
*Brent, Francis (1816–1903),   May 1848 onwards
   London, then Liverpool, then Folkestone;
   clerk, H.M. Customs
*Brewer, James Alexander (1818–1886),   Dec. 1837–51 or later (LS)
   Reigate; postmaster
*Bromfield, William Arnold, M.D. (1801–1851),   fl. 1843–8
   Ryde; independent
Bromley, Charles Nelson (1817–1855),   1840/41–? (LS)
   Stone; general practitioner
Broome, Christopher Edmund (1812–1886),   Apr. 1845–?
   Bristol; independent
*Brown, Edwin (1819–1876),   Nov. 1849–?50
   Burton-on-Trent; bank manager
*Brown, Isaac (1803–1895),   July 1838–?48 (LS)
   Hitchin; schoolmaster
Brown(e), William,   Oct. 1848–?
   (? of Devonport; naval officer)

*Buckland, Revd Samuel (1817–1900),   1839–?41
   Faversham; ordinand, then domestic chaplain
Buckley, Richard Wilson (1822–1875),   Nov. 1846–?
   London; solicitor
*Buckman, James (1814–1884),   July 1837–?44 (LS)
   Cheltenham; chemist and druggist
     (then ironmonger?)
*Bull, Henry (1816?–post 1859),   Mar. 1846–?8
   Godalming; architect and surveyor
Burke, W. J.,   Feb. 1849–?
   Kilbride, Co. Wicklow (? William Joseph
     Burke (1825–1895), Dublin; barrister)
Burnet, Thomas (1797?–1854),   Apr. 1847–?54
   Newcastle upon Tyne; nailmaker
*Butler, Revd Thomas (1806–1886),   fl. 1842–53
   Langar, Notts.; parish priest
[Button, E. H.,   fl. 1842
   London]
[Carrroll, Isaac (1828–1880),   fl. 1850
   Cork; timber merchant]
Carter, James (1797?–1855),   1839–?
   London; seedsman
*Carter, Revd Thomas Garden (1817–1885),   Apr. 1851–?3
   Standon, Herts.; parish priest
Caspary, Johann Xaver Robert, Ph.D. (1818–1887),   Apr. 1850–?1
   Cringleford, Norfolk, then London;
     private tutor
[Castle, Robert (1768?–1850),   fl. 1843
   Twickenham; botanic garden manager]
[*Chadwick, George,   fl. 1852–3
   Accrington]
*Chambers, Gardner (1799–1878),   fl. 1853
   Egremont, Cumb.; grocer
Charlesworth, Edward (1813–1893),   Nov. 1836–?40 (C)
   London; freelance naturalist
Chatterley, William Maddox Foote (1817–1858),   Nov. 1836–? (C; *Secretary*)
   London; science lecturer and medical student
Children, John George (1777–1852),   Aug. 1838–? (C; *Vice-Pres.*)
   London and Halstead, Kent;
     museum taxonomist
*Churchill, George Cheetham (1822–1906),   Oct. 1850–?
   Manchester; solicitor
*Clark, Thomas (1793–1864),   fl. 1843–4; 1849–?51
   Bridgewater; grocer and seedsman
Clarke, Benjamin (1813–1890),   fl. 1844–7
   London; independent

Coleman, Revd William Higgins (1816?–1863), fl. 1843–9
    Hertford, then Ashby-de-la-Zouch;
    schoolmaster
Comfield, Thomas, July 1838–?
    Cheltenham; silversmith and optician
*Conway, Charles (1797?–1860), Dec. 1836–?42 (LS)
    Pontnewydd, Mon.; tin-plate manufacturer
Cooke, John Charles (1811–1858), fl. 1844–9 (C)
    London; medical lecturer and coach
Cooke, Robert Thomas Elsan (c. 1826–c. 1899), Sept. 1851–?
    Scarborough; general practitioner
*Cooper, Daniel (1817?–1842), Nov. 1836–Nov. 1842 (C;
    London; freelance naturalist, then army   *Curator*)
    surgeon
*Cooper, George (1792–1877), Nov. 1837–50 or later (C)
    Brentford; general practitioner
Cooper, Thomas Henry (1813–1881), July 1846–?
    London; general practitioner
Coppin, John (1821–1891), May 1845–? (C)
    London; barrister
Corder, Thomas (1812–1874), July 1838–? (LS)
    Chelmsford, then Adelaide; miller
*Cowell, Matthew Henry (1808–1866), Aug. 1838–44 (LS)
    Faversham; brewer
Crawley, Robert Emans (1819?–1897), Mar. 1837–?
    London; schoolboy
[Cresswell, Revd Richard (1815–1882), fl. 1846
    Salcombe; parish priest]
*Croall, Alexander (1809–1885), fl. 1841–4 (LS)
    Montrose; schoolmaster
*Crotch, Revd William Robert (1799–1877), fl. 1844–53
    Taunton; hospital chaplain, then schoolmaster
Crouch, Revd James Frederick (1809–1888), May 1847–?
    Oxford; don
*Cumming, William (1811–post 1861), 1841?–4
    Audley End, Essex, then Cambridge; gardener,
    then nurseryman
Davies, William St George, M.D. (1786–1882), July 1847–?
    Brighton; physician
Davis, Frederick (1816–c. 1880), Jan. 1848–?
    Lindfield, Sussex; general practitioner
Davis, Robert (1812–post 1847), Nov. 1836–?; May 1847–?
    London; dyer
Dean, James, Mar. 1837–?
    London

*Dennes, George Edgar (1817–post 1858), London; solicitor — Nov. 1836 onwards (C; *Secretary*)

Dennison, Mrs, London — Nov. 1836–?

*Dewar, Andrew (1792?–1870), Dunfermline; general practitioner — fl. 1844–51

*Dickie, George, M.D. (1812–1881), Aberdeen; university teacher — 1839–?49

Dickinson, James, London — Apr. 1849–?

*Dickinson, Joseph, M.D. (1805?–1865), Liverpool; physician and medical lecturer — fl. 1841; Apr. 1846–50 or later

Dickson, Joseph, M.D. (1819–1874), St Helier; physician — Aug. 1849–?

Dorrington, John (1815?–post 1861), Linton, Cambs.; schoolmaster — Feb. 1848–?

*Doubleday, Edward (1811–1849), London; museum taxonomist — 1842?–9 (C; *Vice-Pres.*)

*Doubleday, Henry (1808–1875), Epping; grocer — fl. 1843

Douglas, Francis, M.D. (1815–1886), Kelso; physician — 1842?–5

*Douglas, Revd Robert Cooper (1823–1887), Wolverhampton, then Stafford; parish priest — Feb. 1847 onwards

Duthoit, Thomas James (1828–1858), London; medical student — Feb. 1849–?53

*Dutton, Thomas, Bath; ?gardener (?Thomas Dutton, gardener, 1800?–1876) — July 1850–?

*Dyke, Thomas Jones (1816–1900), Merthyr Tydfil; general practitioner — Nov. 1838–? (LS)

Dyson, Revd Francis (1819?–1887), Cricklade; parish priest — Feb. 1849–?

*Eddison, Booth (1809–1859), Nottingham; general practitioner — Oct. 1846–?

*Embleton, Robert Castles (1806–1877), Alnwick; general practitioner — fl. 1843–9

Evans, William, Llanrwst, Denbighshire — Nov. 1849–?

Evans, Miss, Coventry (?Mary Ann Evans, 1819–1880, the novelist 'George Eliot') — Oct. 1850–?

Ewing, John William (1815?–1868), Norwich; nurseryman and florist — fl. 1843 (LS)

Farenden, Miss Emma (1800?–1880),   July 1845–?
  London; schoolteacher
Farran, Charles, M.D. (1790–1861),   Apr. 1847–?
  Stradbally, Co. Waterford; independent
*Fielding, Henry Borron (1805–1851),   Mar. 1838–43 or later (LS)
  Lancaster; independent
*Fitt, George (1809?–1893),   fl. 1844; Mar. 1846–?8
  Great Yarmouth; accounts clerk
*Flower, Thomas Bruges (1817–1899),   fl. 1840
  London University and Bath; medical student,
    then independent
*Foggitt, William (1835–1917),   1851/2 onwards
  Thirsk; chemist and druggist
*Fordham, Henry (1803–1894),   1840/41–9 or later (LS)
  Royston; banker
*Foster, Miss Susanna (1814–1847),   1839/40–?42
  Luton; independent
Francis, George William (1800–1865),   1842?–9 (C)
  London; schoolmaster, then freelance writer
*Freeman, Joseph (1813–1907),   Nov. 1836 onwards (C)
  London; general practitioner
[Freeman, Samuel (c. 1811–post 1841),   fl. 1842
  Birmingham; gardener]
*French, Joseph Barnabas (1825–1911),   Jan. 1848–?52
  London; chemist and druggist, then clerk
Frodsham, William,   Jan. 1837–?
  London (? chronometer maker, 1819?–1901)
Gace, Revd Frederic Aubert (1812–1902),   Aug. 1848–?
  London; medical student
*Gardiner, William (1808–1852),   1837–?44
  Dundee; umbrella repairer
[Garnett, Revd Richard (1789–1850),   fl. 1843
  London; museum curator]
Gawler, Mrs Ann (1791–1844),   Nov. 1836–?40
  London; housewife
*Geldart, Herbert Decimus (1831–1902),   fl. 1851–3
  Norwich; wine merchant
Gem, Harvey (1806?–1887),   Mar. 1837–45
  London; solicitor
Gerard, Adam (1801?–1849),   Nov. 1836–49 (C)
  London; wine and general merchant
*Gibson, George Stacey (1818–1883),   fl. 1843–50
  Saffron Walden; banker

*Gibson, Jabez Marriage (1822–1877),   fl. 1844
    Coggeshall, Essex; farmer

*Godley, William (1805?–1862),   Aug. 1848–52 or later
    Wallingford; corn dealer

Goodlad, Joe (1823?–post 1851),   1842/3–? (LS)
    Bury, Lancs.; medical assistant

*Goulding, Francis Henry (1826?–1898),   Apr. 1847–?50
    Plymouth; jeweller and optician

Gourlie, William (1815–1856),   Apr. 1851–?
    Glasgow; merchant

Gray, John   fl. 1846

*Gray, John Edward (1800–1875),   Nov. 1836 onwards (*President*)
    London; museum taxonomist

*Gray, Peter (1818–1899),   1848/50–?
    Dumfries; journalist

[Gray, Samuel Octavus (1828–1902),   fl. 1850
    London; Bank of England official]

Green, John W.,   Aug. 1838–43 or later
    London, then Newcastle upon Tyne; printer

Greene, Thomas Webb (1804–1875),   Nov. 1836–? (C)
    London; barrister

*Greenwood, Alfred (1821–1847),   July 1846–Mar. 1847
    Penzance; independent

*Griffiths, Miss Amelia Elizabeth (1802–1861),   1849?–53 or later
    Torquay; independent

*Gutch, John Wheeley Gough (1809–1862),   1839–43; Apr. 1847–? (LS; C)
    Swansea, then London; general practitioner,
    then Queen's Messenger

*Hailstone, Samuel (1767–1851),   fl. 1841–9
    Bradford; solicitor

*Hall, Thomas Batt (1814–1886),   Nov. 1838–?44 (LS)
    Coggeshall, Essex; silk manufacturer

Hambrough, Albert John (1820–1861),   Apr. 1846–?
    Niton, I.O.W.; landowner

Hancock, Thomas, M.D. (1783–1849),   1836/7–8
    Liverpool; physician

Harbert, William (1776?–1859),   fl. 1841
    Newbury; landowner

*Hardy, John (1817–1884),   Mar. 1847–?
    Manchester; accountant

Hardy, Lieut Robert William Hale (1794?–1871),   July 1845
    Bath; naval officer

Harley, Francis (c. 1797–1849), Loughborough; general practitioner	June 1848–9
*Harris, Henry Barham Mitchell, M.D. (c. 1817–c. 1855), Edinburgh, then Dumfries, then St Helier; physician, then independent	Feb. 1839 onwards
Harton, William Henry (1794?–1839), Cambridge University, then Portsmouth; naval architect	Jan.–Apr. 1839
*Harvey, Miss Elizabeth (1798?–1873), Deal; independent	Nov. 1838–47 or later
Harvey, James	fl. 1840
Hassall, Arthur Hill (1816–1894), Richmond, Surrey, then London; general practitioner	Dec. 1844 onwards (C)
Hawes, Miss Sophia Brunel (d. 1870), London and Mortlake; schoolgirl	fl. 1842
Hawkes, Revd Henry (1805–1886), Southsea; Unitarian minister	Apr. 1846–?
[Hawkins, Mr (?Edward Hawkins, 1780–1867, of the British Museum)]	fl. 1845
Hedger, George Frederick (1821?–1882), London; chronometer maker?	Feb. 1849–?
[Henderson, Mr]	fl. 1848
*Henfrey, Arthur (1819–1859), London; medical student, then professional botanist	1841? onwards (C; *Curator*; *Vice-Pres.*)
Henry, Alexander, M.D. (1819?–1890), London; physician	fl. 1844–5
[*Heys, Abraham (1821–1891), Accrington; textile operative]	fl. 1850–2
*Heysham, Thomas Coulthard (1792–1857), Carlisle; independent	1849/50 onwards
Heywood, James (1810–1897), London; barrister	fl. 1846–7
Hickman, George, (? Great Marlow, Bucks.; former army surgeon)	Nov. 1848–?
[Hill, Miss (?Elizabeth Hill, c. 1760–1850, Barnstaple; independent)]	fl. 1846
[*Hill, Robert Southey (1817–1872), Basingstoke; general practitioner]	fl. 1841–2
*Hind, Revd. William Marsden (1815–1894), Pulverbach, Salop; parish priest	Mar. 1848 onwards
Hodges, Revd Charles Bishop (1796–1864), Sandbach, Ches.; parish priest	June 1845–?

Hodges, Revd Thomas (1791–c. 1882), Sandbach, Ches.; independent	June 1845–?

*Holland, Robert (1829–1893),
    Roy. Agr. Coll., Cirencester and Mobberley,
        Ches.; student, then farmer
	June 1849–?

Hollings, James Francis (1806–1862), Leicester; schoolmaster	fl. 1842–4

*Holman, Henry Martin (1821–1881),
    Reigate, then Hurstpierpoint; medical student,
        then general practitioner
	Sept. 1838–?43 (LS)

Hopkins, Manley (1818–1897), London; loss adjuster in marine insurance	fl. 1837
*Hore, Revd William Strong (1807–1882), Devonport, then Norwich; parish priest	fl. 1842–9
*Hort, Revd Fenton John Anthony (1828–1892), Cambridge; student, then don	Feb. 1848 onwards
Howell, John Warren (1810–1844), Bath; general practitioner	Apr. 1838–Jan. 1844 (LS)
Hubbard, George (1784?–1860), Bury St. Edmunds; general practitioner	1841?–8 (LS)
Hudson, Robert (1802–1883), London; timber merchant	Apr. 1850 onwards (C)
Hussey, James (1808–1879), Salisbury; independent	fl. 1847–9
*Ingall, Thomas (1799?–1862), London; Bank of England official	Mar. 1847–51 or later

*Irvine, Alexander (1793–1873),
    Croydon, then Guildford, then London;
        schoolmaster
	Nov. 1836–?38; Sept. 1847 onwards

*James, Mrs Lucy Jones (1823–1895), Denham, Bucks.; independent	Sept. 1851–3 or later
*Jenner, Revd Henry Lascelles (1820–1898), Chislehurst; parish priest	June 1845–6
Johns, Revd Charles Alexander (1811–1874), London; schoolmaster	fl. 1841–7
Johnson, Charles (1791–1880), London; botanical lecturer and writer	Nov. 1836–8 (C; *Vice-Pres.*)
Jones, Daniel Moore (1826–1847), London; civil servant	Mar.–Dec. 1847
*Jones, Mrs E. M. (or W. D.)	1849–50

Just, John (1797–1852),
    Bury, Lancs.; schoolmaster and botanical
        lecturer
	May 1846–?

Keir, P. F., London; ?gardener	Apr. 1851–?

Kernan, John (1799?–1869),   Nov. 1836–?
    London; florist and seedsman
*Keys, Isaiah Waterloo Nicholson (1818–1890),   Sept. 1851–3 or later
    Devonport; printer and bookseller
*Kingsley, Henry (1817–1885),   Feb. 1839–?41 (C)
    Uxbridge; general practitioner
*Kirk, Thomas (1828–1898),   Feb. 1848 onwards
    Coventry; manager of timber merchants
Knott, John (1806–c. 1863),   Aug. 1838–? (LS)
    Portsea, Hants., then Adelaide;
        general practitioner
Lambergen, S.   Feb. 1847–?
Laughton, John J.,   Jan. 1839–?41 (LS)
    Kingston, Jamaica
*Lawrence, John Zachariah (1829–1870),   Sept. 1848–50 or later (C)
    London; ophthalmic surgeon
*Lawson, George (1827–1895),   fl. 1846–9
    Dundee, then Edinburgh; student,
        then botanical lecturer
*Lees, Edwin (1800–1887),   Dec. 1837–?47 (LS)
    Tewkesbury, then Worcester; independent
*Legge, Miss Louisa Frances Catherine   fl. 1852
    (1818?–1893),
    Wonston, Hants.; independent
*Leighton, Revd William Allport (1805–1889),   fl. 1841; Jan. 1848–50 or later
    Shrewsbury; parish priest, then independent
*Lewis, Waller Augustus (1817–1882),   1836?–8
    London; public health inspector
*Leyland, Roberts (1784–1847),   Mar. 1837–47 (LS)
    Halifax; printer
Lhotsky, Jan F., Ph.D. (1795–1866),   1839–?
    London; journalist
Lindsay, Sir Coutts, Bart. (1823–1913),   Sept. 1851–? (C)
    London and Colinsburgh, Fife; landowner
Lingwood, Charles (?1823–1877),   Mar. 1837–?
    London; schoolboy
*Lloyd, George, M.D. (1804–1889),   Feb. 1845–?
    Warwick; independent
*Lucas, Samuel (1805–1870),   fl. 1841
    Hitchin; brewer
Lukis, Frederick Corbin (1788–1871),   Jan. 1837–? (LS)
    St. Peter Port, Guernsey; independent
*Lumb, William Bingley (1812–1860),   Nov. 1838–? (LS)
    Rochdale; general practitioner

*Luxford, George (1807–1854),   Jan. 1848–52 or later (C)
    London; printer, journalist and
      botanical lecturer
*Lynam, James (1812–1885),   July 1846–52 or later
    Galway; civil engineer
[*Lyon, George Jasper (1816–c. 1862),   fl. 1840–1
    Glasgow; merchant and manufacturer]
   McEwen, George (1820?–1858),   June 1849–?
    Arundel; gardener
   McIntyre, Aeneas, LL.D. (c. 1792–post 1843),   Nov. 1836–8? (C)
    Warley, Essex; schoolmaster or private tutor
[*McIvor, William Graham (d. 1876),   fl. 1846
    Kew; gardener]
   Mackay, James Townsend (1775?–1862),   fl. 1844–6
    Dublin; botanic garden curator
*Macreight, Daniel Chambers, M.D. (1799–1857),   Nov. 1836–43 or later (C;
    London, then Sydney, then St Helier;   *Vice-Pres.*)
      physician
*Mann, Robert James (1817–1886),   July 1846–?
    Buxton, Norfolk; general practitioner and
      science writer
*Marsham, Revd Henry Philip (1817–1892),   June 1847 onwards
    Stratton Strawless, Norfolk; parish priest
   Marten, H. A. (d. 1847)   fl. 1842–7
   Masters, William Arthur (1820?–1847)   Mar.–May 1847
    London; museum curator
*Mateer, William, M.D.,   June 1846 onwards (C)
    Belfast, then London; physician and
      professor of botany
*Maw, George (1832–1912),   Sept. 1847 onwards
    Hastings and R.A.C. Cirencester, then
      Worcester, then Broseley, Salop; schoolboy,
      then agricultural student, then tile manufacturer
*Meehan, Thomas (1826–1901),   fl. 1847
    Kew; gardener
   Meeson, Henry Ashton, M.D. (1816–1846),   Nov. 1838–Oct. 1846 (C)
    Grays, Essex; physician
*Merrick, Josiah (1798–1883),   fl. 1841–4 (LS)
    Manchester; textile commission agent
   Miers, John (1789–1879),   1843? onwards (C; *Vice-Pres.*)
    London; retired civil engineer
[Mill, George Grote (c. 1824–1853),   fl. 1847
    London; clerk, East India Company]
   Mitchell, James, LL.D. (1786?–1844),   Nov. 1836–?8 (C)
    London; actuary
*Mitchell, James, M.D. (1822–1862),   Apr. 1849–?
    Nottingham; general practitioner

*Mitchell, John Garden (1806–1880), London; shipping agent   1839/40–46 (C; *Librarian*)

Mitten, William (1819–1906), London, then Hurstpierpoint; chemist and druggist   fl. 1842–8

*Moggridge, Matthew (1803–1882), Swansea; landowner   fl. 1840–7 (LS)

*Moore, David (1808–1879), Dublin; botanic garden curator   fl. 1842–6

*Moore, John (1804–1868), Leicester; general practitioner   July 1848–?

*Moore, Thomas (1821–1887), London; botanic garden curator and journalist   July 1847 onwards (*Librarian*)

Morgan, Mrs, Southsea   Sept. 1851–?

[Morris, John (1810–1886), London; chemist and druggist]   fl. 1840

[Motley, James (1821?–1859), Llanelly; mining engineer]   fl. 1847

[Moxon, Miss Elizabeth Charlotte (1790–1884), Twickenham; independent]   fl. 1842

Moxon, Revd George Browne (1794–1866), Sandringham; parish priest   fl. 1846

*Munford, Revd George (1794–1871), East Winch, Norfolk; parish priest   fl. 1841 (LS)

*Murley, Revd Charles Hemsted (1822–1873), Cheltenham; schoolboy   1839–?

Nesbitt, Miss S. A. (?Sarah Ann Nesbitt, 1791–1878, finally Chiddingfold, Surrey; independent)   1839–?

*Newnham, Revd William Orde (1825–1893), Farnham; student   Mar. 1845–?

Nichols, Charles   May 1847–?

[*Norman, Alfred Merle (1831–1918), Congresbury, Som.; student]   fl. 1850

Norman, John Paxton (1819–1871), London; barrister   July 1850 (C)

*Notcutt, William Lowndes (1819–1868), Fareham, then Fakenham; chemist and druggist   fl. 1842–5; Jan. 1846 onwards

*Oliver, Daniel (1830–1916), Newcastle upon Tyne; grocer   Mar. 1847 onwards

Ord, William (c. 1806–post 1855), London; general practitioner   Nov. 1838–?9

Orpen, Thomas Hungerford, M.D. (1810–post 1863); independent	Nov. 1846–?
*Paine, William Dunkley (1810?–1887), Reigate; commodity broker	Jan. 1837–43 or later
*Palmer, Edward, M.D. (1817?–1897), London; general practitioner, then superintendent of asylum	fl. 1845–52 (C)
[Pamplin, William (1806–1899), London; bookseller, publisher and herbarium agent]	fl. 1846
*Parish, Revd Charles Samuel Pollock (1822–1897) West Hatch, Som.; parish priest	Dec. 1844–9 or later
Parkin, John (1801–1887), London; general practitioner	May 1847–?8 (C)
Pascoe, Francis Polkinghorne (1813–1893), St Austell, then London; independent	July 1845 onwards (C)
Pearson, John (?John Pearson, 1803?–1882), Bewdley, Worcs.; gardener	1842?–5
*Perry, Miss Charlotte (1806–1882), Haslemere; independent	Nov. 1836–44
Perry, William Groves (1796–1863), Warwick; bookseller	fl. 1842–6 (LS)
Phillips, Edward, M.D. (1816–1868), Coventry; general practitioner	Feb. 1848–?
Pollard, Charles Frederick (1817–1867), London; general practitioner	Nov. 1836–?
*Prentice, Charles Brightly (1820–1894), Cheltenham; general practitioner	Feb. 1847–50 or later
*Price, John (1803–1887), Denbigh; schoolmaster	June 1848–?
*Purchas, William Henry (1823–1903), Ross-on-Wye; wine merchant	May 1850 onwards
*Ranking, Robert (1784?–1867), Hastings; general practitioner	Jan. 1837 onwards (LS)
Ranking, William Harcourt (1814–1867), London; medical student	1837–8 (C)
*Ray, John, Epping	fl. 1842–53
*Reece, George (1807–1887), Worcester; museum curator and librarian	fl. 1849
*Reynolds, John (1791–1868), London; schoolmaster	Nov. 1836 onwards (*Treasurer*)
Rich, George Montgomery (1819?–post 1857), London; bookseller	Nov. 1836–50 or later (*Librarian*)

Rich, James Moore (c. 1821–post 1850), London, then Minorca; independent or – as later – naval engineer	Nov. 1836–49 or later (C; LS)
*Riley, John (c. 1796–1846), Papplewick, Notts.; land agent	May 1838–Dec. 1846
*Riley, Mrs Margaretta (1804–1899), Papplewick, Notts.; independent	1839–?40; Apr. 1847–?
Robinson, J. (?Joseph Robinson, 1811–1890, London; gardener)	fl. 1845–6
*Robinson, Mrs Mary Anne Thomasin (1775–1847), Fareham; independent	Sept. 1845–Sept. 1847
*Roby, John (1793–1850), Rochdale, then Great Malvern; banker	Oct. 1846–June 1850
Rock, William Frederick (1802–1890), London; stationer	Nov. 1836–?
*Rogers, Edward (?1817–1866), London; general practitioner	Mar. 1837–?
*Rogers, George London	1839/44–5
Rogers, Jasper Wheeler (fl. 1841–60), London; civil engineer	June 1847 onwards (C)
[Roods, Miss]	fl. 1842
*Rutherford, Andrew (c. 1797–1854), Kingussie, Inverness-shire; schoolmaster	fl. 1838–41
*Rylands, Thomas Glazebrook (1818–1900), Warrington; wire manufacturer	Dec. 1838–? (LS)
*Salmon, John Drew (1802?–1859), Godalming, then London; ice merchant	fl. 1843 onwards (LS; C)
*Salter, John William (1820–1869), London; palaeontologist	Jan. 1848–?
*Salter, Thomas Bell, M.D. (1814–1858), Poole, then Ryde; physician	Nov. 1836–47 or later (C; LS)
*Sandys, Revd George William (1812?–1848), Woolwich; parish priest	Mar. 1846–May 1848
*Sansom, Thomas (1818–1872), London, then Liverpool; clerk, H.M. Customs	Nov. 1837–51 or later (*Librarian*; LS)
Saunders, Bernard (1792–1859), St Helier; nurseryman and seedsman	Feb. 1845–?
*Sawbridge, Miss Frances (1807–1880), Bath; independent	fl. 1845
*Seeley, John (1806?–post 1845), Thames Ditton; bookseller and printer	Feb. 1837–45

Semple, Robert Hunter, M.D. (1814–1891), London; physician — July 1848–9

Sidebotham, Joseph (1824–1885), Manchester; calico printer — fl. 1842–4; Aug. 1846–8

Sidney, Marlow John Francis (1774–1859), Morpeth; landowner — Apr. 1838–46 or later (LS)

Sim, Robert (1791–1878), Foots Cray, Kent; nurseryman — fl. 1847

Simon, John (1816–1904), London; medical student, then houseman in hospital — Jan. 1837–?41 (C)

*Simpson, Samuel (1803?–1881), Lancaster; solicitor — 1839–?44 (LS)

Sly, Stephen (fl. 1838–46), London; engraver — Nov. 1836–?

*Smith, Revd Gerard Edwards (1804–1881), North Marden, Sussex, then Cantley, Yorks., then Chester; parish priest — fl. 1841–7

Smith, John Peter George (1818?–1909), Liverpool; general merchant — June 1849

*Southby, Anthony, M.D. (1800?–1883), Bulford, Wilts.; independent — Oct. 1846–50 or later

Sowerby, Charles Edward (1795–1842), London; bookseller — Nov. 1836–?7 (C)

Spicer, Thomas Trevethan (1814?–1861), London; ?, then pupil barrister — Feb. 1845–?

Stainforth, Revd Francis John (1797–1866), London; parish priest — May 1837–?

*Stark, Robert Mackenzie (1815–1873), Edinburgh; nurseryman — 1839–?

*Stephens, Henry Oxley (1816–1881), Bristol; general practitioner — fl. 1840–3; Feb. 1845 onwards

*Stevens, Charles Abbot (1817–1908), Cambridge University and Rochester; student — Nov. 1838–?40

Stewart, Zachariah Robert (1811–1865), Torquay; general practitioner — Nov. 1848–?

*Stock, Daniel (1797?–1873), Bungay, Suffolk; bookseller and stationer — May 1838–51 or later (LS)

*Storey, John (1801–1859), Newcastle upon Tyne; schoolmaster — fl. 1844; Jan. 1846–50 or later

Stovin, Miss Margaret (1756–1846), Chesterfield; independent — 1842?–5

Stowell, Revd Hugh Ashworth (1830–1886),   fl. 1855
    Faversham; parish priest

*Streeten, Robert James Nicholl, M.D. (1800–1849),   Mar. 1838–46/9 (LS)
    Worcester; physician

*Syme, John Thomas Irvine (1822–1888),   Apr. 1851 onwards (*Curator*; C)
    London; botanical lecturer and curator

*Tatham, John (1793–1875),   1839–49 or later (LS)
    Settle; chemist and druggist

*Taylor, Henry (1813–post 1855),   July 1846–50 or later
    London, then Godalming; chemist and druggist

Taylor, William (1806–1848?),   Mar. 1847–8
    London; grocer

Tebbitt, Walter (1827?–1893),   Oct. 1848–?
    London; importer and mother-of-pearl dealer

[*Thompson, John (1778?–1866),   fl. 1842–6
    Ridley, Northumb.; miller]

Thomson, Spencer (1817–1886),   1842–9 (LS)
    Burton-on-Trent; general practitioner

*Thwaites, George Henry Kendrick (1812–1882),   Oct. 1838–49 (LS)
    Bristol; accountant, then botanical lecturer

Townsend, Frederick (1822–1905),   Feb. 1847–?
    Cambridge University and Ilmington,
       Warwicks.; student

[*Trimmer, Revd Kirby (1804–1887),   fl. 1847
    Norwich; parish priest]

Turner, Thomas (1792?–1873),   Mar. 1848–?
    London; general practitioner

*Twining, Thomas (1806–1895),   1839/40–47 or later (C)
    Twickenham; independent

[Unwin, William Charles (1811–1887),   fl. 1850
    Lewes; schoolmaster]

*Varenne, Ezekiel George (1811–1887),   Aug. 1847 onwards
    Kelvedon, Essex; general practitioner

Walker, D. (?David Walker, fl. 1844–58,   fl. 1846
    Colchester; schoolmaster)

*Wallis, Arthur (1816–1856),   Feb. 1837–?42 (LS)
    Chelmsford; printer

*Ward, James (1802–1873),   fl. 1842; Aug. 1847 onwards (LS)
    Richmond, Yorks.; chemist and oil merchant

Wardale, Benjamin Dornale (1770?–1854),   1839/41 onwards
    King's Lynn; retired naval officer

*Warner, Septimus (1821–1911),   1839/40–6 (LS)
    Hoddesdon, Herts.; banker

*Watkins, Burton Mounsher (1816–1892),   fl. 1844–7
    Ross-on-Wye; relieving officer

 Watkins, Mrs Sophia Louisa Henrietta Lloyd   Apr. 1850–?
    (1802?–1851), Brecon; independent

*Watson, Hewett Cottrell (1804–1881),   1839/40 onwards (*Vice-Pres.*)
    Thames Ditton; independent

 Webb, Revd Robert Holden (1806–1880),   Aug. 1848–?
    Essendon, Herts.; parish priest

*West, William James (1794?–1848),   1839–45/8 (LS)
    Tunbridge Wells; general practitioner

*Westcombe, Thomas (1815–1893),   fl. 1845–50
    Worcester; assistant in tallow chandler's shop,
      then counting-house clerk

*White, Adam (1817–1879),   Feb. 1839 onwards
    London; museum taxonomist

 White, Francis Isaiah, M.D. (1815–1898),   Feb. 1839–43 or later (LS)
    Edinburgh, then Perth; physician

 White, William Henry (1777–1866),   Nov. 1836–42 or later (C)
    London; civil servant

 Whitla, Francis (1783–1855),   fl. 1846–7
    Belfast and Dublin; solicitor

*Whittaker, Joseph (1812–1894),   Feb. 1847 onwards
    Morley, Derbyshire; schoolmaster

 Wilkins, Miss Charlotte (1804–1876),   Nov. 1847–?
    Bristol; independent

 Wilkinson, Joseph Sheldon (1808–1861),   Nov. 1836–?40 (C)
    Syston, Leics.; medical student

[*Willmott, Abraham Taylor (1808?–1890),   fl. 1852
    Ross-on-Wye; general practitioner]

 Willshire, William Hughes, M.D. (1816–1899),   Nov. 1838–?47 (C)
    London; physician and botanical lecturer

*Wilson, James Hewetson (1826–1850),   Apr. 1846–Nov. 1850 (C)
    Oxford University and Worth, Sussex,
      then London; student

 Wilson, Miss Martha,   Aug. 1849–?
    Belfast

*Wing, William Edward (1827–1855),   Sept. 1851–5
    London; artist

*Withers, Robert (1814?–1855),   1849?–55
    Bath; draper

*Wood, John Bland, M.D. (1813–1890), Manchester; physician   fl. 1842–8

Woods, Robert Carr (1816–1869?), London; meteorological instrument maker   Nov. 1836–?38

Woodward, Samuel Pickworth (1821–1865), London, then Cirencester, then London; museum curator, then professor of natural history, then palaeontologist   1839/40–49 or later (C; *Curator*)

Woollett, John (1821?–1898), London; barrister   Aug. 1846 onwards (C)

*Worsley, Miss Anna, later Mrs Russell (1807–1876), Bristol; independent   1839/41 onwards

Wynne, John Arthur (1801–1865), Sligo; politician   May 1846–9 or later

*Young, James Forbes, M.D. (1796–1860), London; general practitioner   Nov. 1836–?45 (C)

APPENDIX THREE

Officers 1836–1986

Abbreviations used: † died in office, * salaried, ‡ acting only

BOTANICAL SOCIETY OF LONDON

President 1836–56 J. E. Gray

Treasurer 1836–56 J. Reynolds

Vice-Presidents		*Secretary*		*Curator*	
1836–8	C. Johnson	1836–7	W. M. Chatterley	1836–41	D. Cooper
1836–40	Dr D. C. Macreight	1837–56	G. E. Dennes	1851	J. Geiger*
1838–44	J. G. Children			1841–4	A. Henfrey*
1840–4	H. C. Watson	*Librarian*		1848	S. P. Woodward*
1844–?48	Dr F. Bossey	1841–5	T. Sansom	1851–6	J. T. Syme*
1844–56	J. Miers	1845–6	J. G. Mitchell		
1844–9	E. Doubleday†	1846–8	G. M. Rich		
1850–56	A. Henfrey	1848–56	T. Moore		

BOTANICAL EXCHANGE CLUB

Secretary		*Curator*	
1856–66	R. D. Carter	1857–68	J. G. Baker
1868–79	J. G. Baker	1857–?64	J. H. Davies
1868–71	H. Trimen	1865–6	W. Foggitt
1879–1903	C. Bailey	1868–75	J. T. Syme
1903–32	G. C. Druce†	1874–5	J. F. Duthie
		1875–8	T. R. Archer Briggs
		1878–9	A. R. Pryor

Post-1932 Reconstitution

Chairman

1932–47	Rt Hon H. T. Baker
1947–8	J. S. L. Gilmour

Vice-Chairman

1947	A. H. G. Alston

General Secretary

1932–6	W. H. Pearsall†
1936–9	J. F. G. Chapple
(1939–41	E. F. Druce†)‡
(1941–5	A. J. Wilmott)‡
1945–7	J. F. G. Chapple

Treasurer

1932–4	T. J.† & Mrs G. Foggitt
1934–7	Sir Roger Curtis
1937–41	F. Druce†
1941–7	J. E. Lousley

Editors

1936–41	P. M. Hall†
1941–7	A. J. Wilmott
1941–7	E. C. Wallace

BOTANICAL SOCIETY OF THE BRITISH ISLES

President

1948–51	J. S. L. Gilmour
1951–5	Canon C. E. Raven
1955–7	Dr G. Taylor
1957–61	Prof T. G. Tutin
1961–5	J. E. Lousley
1965–6	Dr E. F. Warburg†
1967–9	Dr J. G. Dony
1969–71	E. Milne-Redhead
1971–3	D. McClintock
1973–5	Dr S. M. Walters
1975–7	E. L. Swann
1977–9	Prof D. H. Valentine
1979–81	R. W. David
1981–3	Prof J. P. M. Brenan
1983–5	J. F. M. Cannon
1985–	D. E. Allen

Vice-Presidents

1948–51	A. H. G. Alston
1950	Miss M. S. Campbell
1951–5	Dr R. W. Butcher
1951–3	J. F. G. Chapple
1951–5	J. S. L. Gilmour
1952–6	Prof T. G. Tutin
1954–6	N. D. Simpson
1955–9	G. M. Ash
1955–9	Prof D. A. Webb
1956–8	N. Y. Sandwith
1956–60	J. E. Lousley
1958–62	E. Milne-Redhead
1959–64	J. E. Dandy
1959	Dr R. C. L. Burges†
1960–4	Dr E. F. Warburg
1960–4	Miss C. M. Rob
1962–6	Dr S. M. Walters
1963–6	Dr C. E. Hubbard
1964–7	R. Mackechnie
1964–7	Dr J. G. Dony
1966–9	E. Milne-Redhead

Vice-Presidents

1967–9	Miss U. K. Duncan
1967–71	Prof J. G. Hawkes
1968–72	Prof D. H. Valentine
1969–73	Dr F. H. Perring
1969–73	J. E. Lousley
1971–5	Prof J. Heslop-Harrison
1972–6	J. C. Gardiner
1973–7	Dr W. T. Stearn
1973–7	Mrs H. R. H. Vaughan
1975–9	Dr C. T. Prime
1976–80	Mrs B. H. S. Russell
1977–81	Prof J. P. M. Brenan
1977–81	J. F. M. Cannon
1979–83	D. H. Kent
1980–4	P. C. Hall
1981–5	R. W. David
1981–5	Dr S. M. Walters
1983–	Prof. D. A. Webb
1984–	E. C. Wallace
1985–	Dr N. K. B. Robson

General Secretary
1947–50	Miss M. S. Campbell
1950–6	J. E. Lousley
1956–64	Dr J. G. Dony
1964–7	E. B. Bangerter
1967–9	D. E. Allen
1969–72	Dr I. K. Ferguson
1972–	Mrs M. Briggs

Meetings Secretary
1953–6	Dr J. G. Dony
1956–64	Dr H. J. M. Bowen
1964–72	Mrs M. Briggs
1972–5	Miss G. Tuck (later Mrs Beckett)
1975–6	Dr J. T. Williams
1976–80	Mrs J. M. Mullin
1980–5	Miss J. Martin (later Mrs Robertson)

Treasurer
1947–50	J. E. Lousley
1950–8	E. L. Swann
1958–71	J. C. Gardiner
1971–	M. Walpole

Excursions Secretary
1947–9	Miss M. S. Campbell

Field Secretary
1949–53	Dr J. G. Dony
1959–67	P. C. Hall
1975–6	Mrs G. Beckett
1976–82	Miss L. Farrell
1983–4	J. N. B. Milton†
1985–	R. Smith[1]

Junior Activities Secretary
1959–61	Miss M. E. Bradshaw
1961–5	P. F. Hunt

Editors
1947–60	Dr E. F. Warburg
1953–66	D. H. Kent
1960–71	Dr M. C. F. Proctor
1965–71	E. F. Greenwood
1967–	Dr N. K. B. Robson
1968–9	Dr C. D. K. Cook
1969–79	Dr G. Halliday
1971–83	Dr C. A. Stace
1977–84	Dr S. M. Eden
1977–83	Dr D. L. Wigston
1982–	Dr R. J. Gornall
1983–	Dr J. R. Akeroyd
1984–	Dr B. S. Rushton
1985–	E. D. Wiggins

Membership Secretary
1969–75	Mrs C. M. Dony
1975–81	Mrs R. M. Hamilton

[1] By a rule change in May 1985 these office-holders ceased to have officer status.

APPENDIX FOUR

Principal BSBI Conferences

1948	"The Study of Critical British Groups"
1950	"Aims and Methods in the Study of the Distribution of British Plants"
1952	"The Changing Flora of Britain"
1954	"The Species Concept in its relation to the British Flora"
1956	"Progress in the Study of the British Flora"
1959	"A Darwin Centenary"
1961	"Local Floras"
1963	"The Conservation of the British Flora"
1965	"Reproductive Biology and Taxonomy of Vascular Plants"
*1967	"Modern Methods in Taxonomy"
1969	"The Flora of a Changing Britain"
*1971	"Taxonomy and Phytogeography of Higher Plants in relation to Evolution"
1972	"Plants Wild and Cultivated" (with the Royal Horticultural Society)
1973	Symposium on the British Oak
*1974	"European Floristic and Taxonomic Studies"
*1977	"The Pollination of Flowers by Insects"
1978	Aquatic and Marsh Plants Symposium
1979	"Recent Advances in the Study of the British Flora"
*1980	"Biological Aspects of Rare Plant Conservation"
1980	Fern Symposium (with the British Pteridological Society)
1984	"Archaeology and the Flora of the British Isles" (with the Association for Environmental Archaeology)
1985	"Recording Critical Groups in the Flora of the British Isles" (with the Biology Curators Group)

* In association with the Linnean Society of London

INDEX

Adelaide 49
Alchemilla 140
Aldermaston 157
Allman, G. J. 47
Almquist 112
Alps 53
Amazon 114
America, North 18
American Lady's Tresses 157
Anglesey 107
Annals and Magazine of Natural History 9
Annals of Scottish Natural History 96
Apera interrupta 33
Aphanes microcarpa 141
Apothecaries Act 6
Arbroath 111
Ashmolean NHS 105
Association
    British 64
    Geologists' 83, 169
    Linnaea 83
Athenaeum, Portsmouth 64
Atlas of the British Flora 153, 158, 160
Atriplex 134
Australia 19, 25, 49, 61
Austria 112
Ayres, P. B. 18, 60
Azores 33

Babington, C. C. 9, 26, 31–2, 35, 58, 72, 78–9, 88
Backhouse, James 74
Bacon, Miss Gertrude 119
Bagnall, J. E. 78
Bailey, Charles 73, 80–5, 87–9, 91–3, 96, 105, 110, 160
Baker, H. T. 119, 121
Baker, J. G. 56, 61, 69–77, 79–80, 82, 88, 91, 133, 138–9
Balfour, J. H. 30, 32, 35, 58
Balmuto 55
Barlow, T. W. 50
Barnard, F. 49
Barrington, R. M. 101
Bean, W. 44
BEC, *see* Club, Botanical Exchange
Beeby, W. H. 78
Beesley, Thomas 43
Belgium 19
Bennett, Arthur 101, 113

Bentham, George 34, 77, 88, 109, 125
Berkshire 157
Berwickshire 19
Betula 140
Birches 140
Blackpool 81
Blow, T. B. 89
Bloxam, Andrew 44
Blue Heath 108
Bonplandia 72
Bootham School 69
Borneo 50
Borrer, William 75
Botanical Gazette 30
'Botanist, The' 164
Box Hill 20, 33
Braemar 30
Brambles 32, 85, 110, 112
Brean Down 151
Breckland 33
Bree, W. T. 44
Brewer, J. A. 53
Briggs, T. R. Archer 80
Bristol 102
British
    Association 64
    Guiana 19
    Isles 4, 8–9, 45, 107, 149, 152, 154, 157, 171–2
British
    Botanist 71
    Herbaria 152
    Plant List 126
Britten, James 95, 98, 101, 122
Britton, C. E. 112
Brocas, F. Y. 60
Bromfield, W. A. 32
Brontë, Branwell 50
BSBI, *see* Society, Botanical of the British Isles
BSBI News 168
BSL, *see* Society, Botanical of London
Buchan, John 119
Buckle, O. 151
Buckman, J. 47, 64
Buncle & Co. 142
Butcher, R. W. 120
Butler, Samuel 50

Caernarvonshire 107
Caithness 56, 159

Calamint 32
Calamintha sylvatica 32
Calcutta 50
Cambridge 3, 6, 9, 97–8, 102, 125, 148–9, 154
Cambridge British Flora 97, 102–3
Cambridgeshire 108
Campbell, Miss May Sherwood 138–40, 143–7, 151–2
Canada 49
Capsella 112
Carex
   appropinquata 33
   montana 33
   paupercula 32
Cardinal Numbers, The 40
Carpenter, Margaret 37
Caspary, J. X. R. 47
Castle of Mey 159
Centaurea 112
Chadwick, Edwin 42, 54
Chapple, John 116, 118–9, 123, 126, 139, 142, 144–6
Charlesworth, Edward 14
Chatterley, William 11, 12, 14–5, 22, 29
Chelsea 16
   Physic Garden 53
Cheshire 25
Churchill, G. C. 53
Cirencester 64, 80
Clapham, A. R. 153–4
Club
   Botanical Exchange 44, 67, 73, 77, 83, 85, 89, 96, 98–9, 101, 104, 106–7, 109–11, 113–20, 123–9, 133–4, 137–8, 147, 150, 171
   Botanical Locality Record 89
   Entomological 5, 13, 46
   Foreign Exchange 56
   Moss Exchange 83–4
   Phytological 83
   Quekett Microscopical 73
   Ray 9
   Thirsk 50, 72–3, 119
   Watson 83, 88–9, 91, 105, 108, 121, 129, 133
Colgan, Nathaniel 89
Comital Flora of the British Isles 113, 124, 127
Conservancy, Nature 151, 154, 158, 166, 168
Cooke, M. C. 72
Cooper
   Daniel 6–7, 11–12, 14–15, 17, 20, 22, 24, 27, 34, 40, 41
   John Thomas 6
Corstosphine, R. H. 111, 118
Cow Green 163
Cranberry 33
Critical Supplement to the Atlas of the British Flora 158
Croall, Alexander 31
Crouch, J. F. 47
Crowfoots, Water 106
Cumming, Linnaeus 99
Cut-leaved Germander 33

Cybele
   Britannica 36, 63, 153
   Hibernica 89

Dactylorchids 108, 125
Dactylorhiza cruenta 141
   incarnata 141
Dandy, J. E. 123, 127, 152
Dardanelles 102
Darlington 74
Dartmoor 157
Darwin, Charles 6
Davies, J. H. 71
Dennes, George Edgar 14, 29, 31, 36–7, 41–3, 45, 53, 57–8, 61
Devon 106
Dick, Robert 56
Dickie, G. 47
Docks 134
Dony
   J. G. 144, 151, 159
   Mrs 159
Drabble, Eric 112
Druce
   Francis 128
   George Claridge 91–6, 98–102, 104–12, 114–21, 123–4, 126–8, 133–5, 137, 144, 171
Dunning, J. W. 87
Duthie, J. F. 79
Duthoit, T. J. 50
Dyer, W. Thiselton 73
Dyke, T. J. 43

Eastwood, Dorothea 152
Edinburgh 6, 8, 9, 15, 25–6, 28, 34–5, 37, 55, 62, 106
Edmondston, Thomas 30–1, 55
Eliot, George 50
English Botany 77, 79, 88, 114
Entomologist's Record 83
Epipactis 141
Equisetum ramosissimum 141
Esslinger Reisegesellschaft 8
Euphrasia 140
Evans
   A. H. 108
   Miss 50
Eyebrights 140

Fife 55, 77
Flora
   Europaea 172
   Metropolitana 7
   of Berkshire 95
   of Hertfordshire 80
   of Middlesex 73
   of Oxfordshire 92
   of Plymouth 80
   of North Yorkshire 74
   of Surrey 118

Flora – cont.
  of West Ross 111
  of West Yorkshire 80
  of the British Isles 153
Foggit
  Mrs 120, 123–5, 127
  T. J. 119
  William 75–6
Forbes, Edward 9, 30
Francis, G. W. 48–9
Freeman, J. 14, 18
French, J. B. 53
Freud, S. 26
Friends, Society of 44
Fryer, Alfred 101

Gardiner, J. C. 160
Geiger, Joseph 30
Gerard, Adam 34
German Travelling Union 8
Germander, Cut-leaved 33
Gilmour, J. S. L. 120, 124–5, 138
Gissing, George 50
Glasgow 6, 15, 30
Glasnevin 64
Gloucestershire 108
Goode, R. H. 121
Graham
  R. 6
  R. A. 142
Grass, Holy 56
Gray
  J. D. 83
  John Edward 4, 13–5, 22, 26, 37, 43–4, 61
  Samuel Frederick 4
Greenwich 30
Greenwood, E. F. 141
Griffiths, Miss A. E. 44
Groves, James 83
Guernsey 151
Guiana, British 19
Guide, The New Botanist's 24

Hall
  P. C. 151
  Patrick Martin 124–8
Hamlet 94
Hampshire 80, 157
Hanbury, F. J. 78, 108
Handbook of the British Flora 77, 88, 125
Hardwicke, Robert 73
Hardy, John 60
Harford, W. A. 102
Harvey, Miss E. 44
Hassall, A. H. 42
Hawkweeds 85
Helleborines 141
Henfrey, Arthur 30, 49, 55, 63–4, 166
Henslow, J. S. 6, 9, 32
Hertfordshire 33

Heysham, T. C. 44
Heywood, James 43
Hiern, W. P. 106
Hierochloe borealis 56
Hochstetter, C. F. 8
Hollings, J. F. 47
Holy Grass 56
Hooker
  J. D. 75, 88
  Sir William 6, 15–16, 27, 30–1, 36, 60, 72
Hopkins, M. 40–1
  Gerard Manley 50
Horsetail 141
Hort, F. J. A. 47
Horwood, A. R. 109
Hull 84
Hunnybun, E. W. 97
Hunt, T. C. 33

India 80
Institution
  Plymouth 64
  Royal, of South Wales 64
Ireland 32–3, 88–9, 141, 149–50, 157, 172
Irish Topographical Botany 89
Irvine, Alexander 13, 36–7, 57, 62, 71–2
Isle
  of Man 152
  of Wight 32

Jersey 107
Johns, C. A. 47
Johnson
  C. 14
  Thomas 3
Journal of Botany 72, 77, 79, 80, 88, 95–6, 105, 113, 140
Just, J. 47

Kent 79
Kent, D. H. 141, 151–2, 159
Kew, Royal Botanic Gardens 6, 16, 25, 33, 72–3, 75–6, 82, 89, 109, 117, 120, 125, 149
Kirk, Thomas 49
Kloos, A. W., Jr 154

Lady's Mantles 140–1
Lambert, A. B. 30
Lambeth 6, 7, 72–3
Lawson, George 49
Lees
  Edwin 18, 36
  F. Arnold 78–9
Leicester 109
Leucojum aestivum 30
Ley, Revd Augustin 91
Lhotsky, J. 19, 49
Limonium binervosum 64
Lincolnshire 3, 141

Linnean Society 3–5, 10, 12, 19, 30, 38, 46, 61–3, 90, 105, 120, 133
Linnaeus 4, 114
Linnell, John 50
Linton
   E. F. 44, 78, 82–4, 87–9, 93, 97
   W. R. 78
List of
   British Plants 96
   British Vascular Plants 152
Liverpool 62
Loddon
   Lily 30
   Pondweed 136
London Catalogue of British Plants 34–5, 60, 69, 95, 152
Loudon, J. C. 14
Lousley, Job Edward 133–5, 137–9, 142, 144–6, 151, 153–4, 159–60
LSA 6
Luxford, George 47, 69
Luzula sylvatica 17

McIntyre, Aeneas 14–15, 40–1
Mackechnie, R. 149
Macreight, D. C. 14, 17, 25
Magazine of Natural History 8, 14
Malta 79
Manchester 81–2, 97, 149
   Literary and Philosophical Society 45
Manual, Babington's 88
Marriage family 44
Marshall, Revd E. S. 92, 96–9, 101, 106, 108, 109
Marsham
   Henry 44
   Robert 44
Martyn, John 3
Masters, W. A. 44
Mateer, W. 47
Medical Dictionary 15
Meeson, H. A. 14
Melampyrum 112
Mertensia maritima 159
Mickleham 20
Miers, John 15
Milium scabrum 141
Milkwort 79
Mill
   G. S. 15
   John Stuart 15, 34
Mills, W. H. 108
Milne-Redhead, E. 138
Minuartia stricta 33
Mirror of Flowers 152
Mistletoe 167
Mitchell, J. 14
Moggridge, Matthew 44
Monks Wood 158
Moore
   David 32, 64, 89
   Thomas 48, 53

More, A. G. 88
Moss, Charles Edward 97–103, 121–2
Motley, J. 50
Museum
   British (Natural History) 4, 13–14, 27, 95, 97–8, 102, 117, 120, 125, 133, 145, 149, 152
   Horniman 60
Museums Act 64

Najas flexilis 33
New Botanist's Guide, The 25
New Student's British Flora 125
New Phytologist 141
New Zealand 49
Newbould, Revd W. W. 73, 75
Newman, Edward 22, 69, 71
Nicholson, George 82
Norman, J. P. 50
Northampton 92
Northamptonshire 92
   NHS 109
Northumberland 32

Oenothera 96
Oliver, Daniel 33
Oraches 134
Orchid
   Marsh 141
   Military 157, 173
   Monkey 108
Orchis
   militaris 157
   simia 108
Orkney 55
Outlines of the Geographical Distribution of British Plants 25
Oxford 91–2, 98, 105, 109, 113–4, 116–9, 125, 139, 144
Oxfordshire 60, 105
Oyster Plant 159, 162

Palmer, Samuel 50
Pamplin, William 8, 47, 71
Parsley Piert 141
Pearsall, William Harrison 109, 118–24, 126–7, 129
Perring, F. H. 154–5, 158
Phrenological Journal 26
Phyllodoce caerulea 108
Phytologist 9, 13, 22, 34, 36–7, 45, 52, 62, 69, 72
   New 141
Poe, Edgar Allan 50
Polygala amarella 79
Pondweeds 123
Portsmouth 125
Potamogeton 123
Praeger, R. Lloyd 89, 171
Prentice, C. B. 49
Price, W. R. 146, 151
Primrose 86
Proceedings of the Botanical Society of London 21, 71

Proceedings of the Botanical Society of the British Isles 141, 151, 159, 160, 167–8
Pryor, A. R. 80
Pugsley, H. W. 105, 129, 138

Quakers 44, 69, 83
Queen Mother, The 159, 161

Ralli Brothers 81, 160
Ramsbottom, J. 120–1
Ray, John 11, 44
Regent's Park 16, 62
Reigate 20
Rendle, A. B. 98, 110
Reynolds, John 14, 40–1, 43, 58
Ribbons, B. W. 149
Rich, Obadiah 18
Riddelsdell, H. J. 44, 112
Riley, Mrs Margaret 18, 45
Rogers
   J. W. 42
   W. Moyle 110
Roman Nettle 95
Ronniger 112
Rosa 74, 112
Roses 32
Roundstone 33
Rousseau, J.-J. 18
Royal
   Agricultural College 80
   Institution 41
Rubi 18
Rumex 134
Rylands, T. G. 43

St Andrews 114
St Brody, G. A. O. 108
Salmon, C. E. 98, 118
Salter, T. Bell 14
Sandwort, Teesdale 33
Sansom, T. 27
Schomburgk, Robert 19
Scirpus sylvaticus 17
Science Gossip 83
Scotland 54, 79, 150
Scully, R. W. 89
Scutellaria hastifolia 33
Sea Lavender, Irish 64
Seemann, Berthold 72, 79
Shaw, George 4
Sherborn, C. Davies 62–3
Sierra Leone 34, 50
Skullcap 33
Sledge, W. A. 120
Sly, S. 48
Smiles, Samuel 56
Smith, Sir James Edward 4, 44
Snowdon 107
Society, Botanical
   of America 170

Society, Botanical – *cont.*
   of London 5, 16, 21, 33, 34, 38, 41, 58, 63, 69, 71–3, 80, 83, 94, 110, 113–4, 120, 145, 152, 166, 170
   of Edinburgh 10, 26, 28, 33–4, 37, 55, 61, 106, 149
   of the British Isles 49, 71, 89, 120, 129, 131, 133, 138–9, 142, 145–9, 158–60, 163, 166, 168–72, 174
   of Ireland 172
British Mineralogical 6
Dublin 64
Entomological 5, 13, 46, 62–3, 87
   Cooperative 83
Geological 21
Geographical 5
Linnean 3–5, 10, 12, 19, 30, 46, 61–3, 90, 105–6, 120, 133
Manchester Literary and Philosophical 45
Metropolitan Improvement 43
Meteorological 5, 14
Microscopical 22
Pharmaceutical 47, 83
Ray 127
Royal 3, 114, 142, 152
   Botanic 16, 62
Statistical 5
Uranian 15
Westminster Reform 43
Wild Flower 107, 124
Zoological 5, 13, 62

Tansley, A. G. 97, 125, 153
Taunton 151
Taylor, G. 123
Teesdale 163–5
   Sandwort 33
Teucrium botrys 33
Thame 60
Thames Ditton 24
Thirsk 69, 71, 80
   Club 50, 72–3, 75, 119
   NHS 75–6
Thompson, H. Stuart 175
Thurso 56
Thymus 112
Times, The 16
Topographical Botany 88, 112, 124, 127, 153
Transactions (of the Botanical Society of Edinburgh) 10
Trimen, Henry 73, 76–9, 82
Tunbridge Wells 33
Turrill, W. B. 109
Tutin, T. G. 153
Twining, Thomas 43, 49

Unio Itineraria 8, 34
Unitarians 44–5
Unwin, W. C. 47
Urtica dodartii 95
Utricularia 127

Uxbridge 20

Vaccinium macrocarpon 33
Vachell, Miss Eleanor 106–7
Varenne, E. G. 43
Victoria amazonica 19
Viola 112

Wade, A. E. 120
Wales 8, 33, 149, 150
Wallace, E. C. 144
Waller, A. R. 88
Walters, S. M. 141, 154, 158, 172
Wapping 4
Warburg, Edmund Frederic 139–41, 153
Ward, F. H. 89, 90, 170
Water Crowfoots 106
Water-lily, Giant 19, 114
Waterfall, Charles 84
Watson
  Club 83, 88, 89, 91, 105, 108, 121, 129, 133
  Hewett Cottrell 7–9, 15, 17, 25–39, 42–3, 48, 50–8, 60–1, 53–4, 69, 71–2, 74–6, 78–9, 81–2, 88–9, 111, 113, 128–9, 134, 150–1, 153, 166
  W. C. R. 112
Watsonia 72, 140–1, 160, 167–8, 170
Wedgwood, Mrs 110

Welch, Mrs B. 151
Westmeath 32
Westrup, A. W. 148
White
  F. Buchanan 78
  Gilbert 44
  J. W. 84, 98
  W. H. 11, 14–5, 19
Whitebeams 140
Wicklow 101
William Wilson 50
Wilmott, Alfred James 102, 120–1, 124–6, 128, 133–4, 138–40, 143–4, 146, 150
Wilson, J. H. 41
Wing, W. 48
Woking 20
Wolley-Dod, A. H. 112
Wood Club-rush 17
Wood-rush, Greater 17
Woodward, S. P. 27, 49

Year Book 141, 144, 146, 151
York 74
Yorkshire 69, 76, 97
Young
  D. P. 141
  James Forbes 6, 7

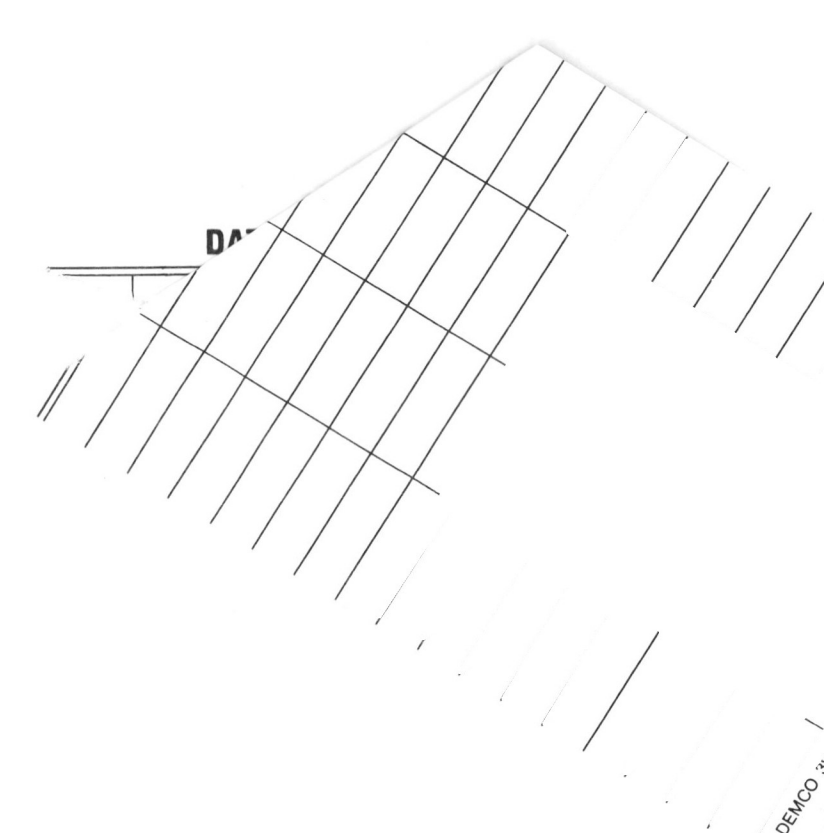